Artificial Intelligence and Robotics in Manufacturing

Artificial intelligence (AI) is revolutionizing the manufacturing industry by automating tasks at a speed, scale and scope that far surpass human capabilities. As AI continues to advance, it is clear that the future of manufacturing will be driven by intelligent systems. These systems have the potential to transform the way products are made, improving efficiency and quality across the board. AI is paving the way for a new era of manufacturing, where innovation and automation go hand in hand.

Artificial Intelligence and Robotics in Manufacturing: A Sustainable Future offers a comprehensive exploration of the scientific and technological advancements in robotics-based manufacturing. This book delves into a variety of topics, including applications in the aerospace, nuclear and medical industries. Emphasizing the use of advanced AI-based robotics, this book highlights recent breakthroughs in manufacturing techniques, specifically focusing on AI applications in fabrication and machining. Additionally, it examines the rapid evolution of AI technology and addresses cost-effective solutions within the manufacturing sector.

Overall, this book provides valuable insights into the intersection of AI and robotics in manufacturing, offering a glimpse into the exciting possibilities for the future of this industry and allowing this book to serve as a valuable source for academics, researchers in manufacturing, computational scientists, mechanical engineers, systems engineers, manufacturing engineers and professionals in industries related to AI and manufacturing.

Innovations in Smart Manufacturing for Long-Term Development and Growth
Series Editors: Atul Babbar, Gursel Alici, Yu Dong, Ankit Sharma

Fabrication Techniques and Machining Methods of Advanced Composite Materials
Vikas Dhawan, Atul Babbar, Inderdeep Singh and Jonathan M. Weaver

Advances in Pre- and Post-Additive Manufacturing Processes: Innovations and Applications
Naveen Mani Tripathi and Ankit Sharma

Handbook of 3D Printing in Pharmaceutics: Innovations and Applications
Prakash Katakam, Ranvijay Kumar, Nishant Ranjan and Atul Babbar

Intelligent Machinery Fault Diagnostics and Prognostics: The Future of Smart Manufacturing
Deepam Goyal, Ankit Sharma and Mohamad Abou Houran

Nanomaterials for Smart Manufacturing: Fundamental Principles, Theory, Case Studies, and Advanced Applications
Dinesh Deshwal, Anil Kumar Narwal, Vivek Srivastava, Atul Babbar and Jonathan M. Weaver

Sensor Technologies and Wireless Communications in Industry 5.0: A Solution for Smart Manufacturing
Edited by Ramkumar K.R., Bhupendra Singh, Lei Pan and Ankit Sharma

Handbook of Smart Manufacturing Technologies: Industrial Engineering in the Digital Age
Edited by Mahendra U. Gaikwad, Ankit Sharma, Muthuramalingam Thangaraj and Sachin Salunkhe

Handbook of Bio-Based Materials for Smart Manufacturing
Edited by Devendra Singh, Naveen Mani Tripathi, Dharminder Singh and Ankit Sharma

Artificial Intelligence and Robotics in Manufacturing: A Sustainable Future
Edited by Ankit D. Oza, Hreetabh Kishore, Ankit Sharma and Atul Babbar

For more information about this series, please visit: www.routledge.com/Innovations-in-Smart-Manufacturing-for-Long-Term-Development-and-Growth/book-series/CRCISMLDG

Artificial Intelligence and Robotics in Manufacturing

A Sustainable Future

Edited by
Ankit D. Oza, Hreetabh Kishore,
Ankit Sharma and Atul Babbar

CRC Press
Taylor & Francis Group
Boca Raton London New York

CRC Press is an imprint of the
Taylor & Francis Group, an **informa** business

Designed cover image: Shutterstock

First edition published 2026
by CRC Press
2385 NW Executive Center Drive, Suite 320, Boca Raton FL 33431

and by CRC Press
4 Park Square, Milton Park, Abingdon, Oxon, OX14 4RN

CRC Press is an imprint of Taylor & Francis Group, LLC

ISBN: 978-1-032-82784-1 (hbk)
ISBN: 978-1-032-82826-8 (pbk)
ISBN: 978-1-003-50649-2 (ebk)

DOI: 10.1201/9781003506492

Typeset in Palatino
by Newgen Publishing UK

The editors dedicate this book to their parents, families and collaborators, whose support laid the foundation for their endeavours. In the end, they also dedicate this book to all the scientific, global collaboration and research communities.

Contents

Preface

The manufacturing sector is rapidly growing with significant integration of artificial intelligence and robotics technologies, and this book, *Artificial Intelligence and Robotics in Manufacturing: A Sustainable Future*, provides an in-depth exploration of the tools and technologies transforming industrial and manufacturing engineering.

It covers a wide range of topics essential for professionals, researchers and engineers aiming to understand this evolving field. The first chapter, "Advanced Manufacturing Techniques of Additive Manufacturing Processes: A State-of-the-Art Review and Future Applications", covers a combination of trendy artificial intelligence (AI) and Industry 4.0 paradigms, Internet of Things (IoT), virtual technology and cloud computing within additive manufacturing systems. Further, a case study is based on machine learning to predict productivity and surface quality in EN19 alloys using electrical discharge machining. Another chapter delves into the implementation of AI and Industry 4.0 in additive manufacturing processes. The book also explores practical applications like damper material along with innovations in AIoT-based manufacturing technologies.

Another chapter, "Green Composite: AI-Based Fabrication, Characterization, Evaluation and Application," focuses on providing a comprehensive evaluation of AI-integrated fabrication approaches compared to conventional methods, emphasizing their impact on material performance and industrial scalability. The book also examines the intersection of IoT-enabled robotic systems on remanufacturing, which applies smart manufacturing innovations. It further discusses technical challenges, limitations and outlines future directions including blockchain integration, edge AI, multi-robot collaboration and standardization of IoT protocols. Other chapters cover topics like AI-based scientometric analysis of post-processing in additively manufactured structure and offer a thorough, methodical analysis of the most current developments in the post-processing of additively built structures.

Overall, this book provides a comprehensive understanding of the manufacturing reshaping the industrial landscape, offering insights into how IoT, machine learning, AI and cloud computing are enhancing manufacturing processes, improving efficiency and tackling emerging sustainable challenges. Also, this volume serves as a useful reference for manufacturing academicians and researchers, manufacturing engineers and professionals in related industries with AI and sustainability.

Acknowledgements

We express our sincere thanks to all the contributors, all those friends and colleagues who have activated our norms to take up a study like this.

We extend our profound gratitude to our institutions, whose unwavering support and dynamic research ecosystems at our universities and institutes have been instrumental in shaping this work.

We owe a deep debt of thanks to our families, whose enduring patience, unwavering encouragement and the nurturing environments they cultivated at home have been the true backbone of this endeavor. Their silent strength and sacrifice have been the invisible yet indispensable force driving us forward.

We also express our sincere appreciation to CRC Press, Taylor & Francis for their professional guidance and support throughout the publication process, ensuring the global reach and impact of this work.

Ankit D. Oza
Hreetabh Kishore
Ankit Sharma
Atul Babbar

Editors

Ankit D. Oza, PhD, is a Research Professor at Parul University in Vadodara, Gujarat, India. He earned a PhD in industrial engineering at Pandit Deendayal Petroleum University (PDPU), Gandhinagar, India. His primary research focuses on non-traditional and hybrid micro-machining processes and their applications, while he also conducts work on AI and IoT-based manufacturing processes. Dr. Oza has published more than 70 articles and over 30 conference papers. He holds more than 45 national and international patents. Dr. Oza regularly delivers expert talks at prestigious universities and colleges across India. He also serves as a reviewer for numerous prestigious international journals.

Hreetabh Kishore, PhD, earned a PhD in mechanical engineering at the Indian Institute of Technology Ropar, Punjab, India. Currently, he works in the Department of Manufacturing and Engineering Technology at Tennessee Tech University in Cookeville, Tennessee, USA. As an enthusiastic educator, Dr. Kishore employs student-focused teaching approaches that provide real-time learning experiences. He has participated in two international conferences with funding from DST, India Travel Grant. His research focuses on AI and IoT-based manufacturing processes and related fields. Dr. Kishore has authored several publications in peer-reviewed journals, with additional papers currently under review, and he has presented and published more than 21 papers at international conferences worldwide. He has also contributed to intellectual property development by filing patents in the advanced engineering domain.

Ankit Sharma, PhD, is the Associate Director (Research) at the Centre for Research Impact and Outcome (CRIO) Department, Chitkara University, Punjab, India, where he also heads the university's international peer-reviewed journals and publications. After earning a PhD in mechanical engineering at the Thapar Institute of Engineering and Technology in Punjab, India, he established himself as a passionate researcher with diverse interests spanning manufacturing, medical devices, modern machining processes, additive manufacturing and tissue engineering. With more than 55 scientific articles published in reputed journals, conferences and books, Dr. Sharma founded the annual ICEMSMCI conference and organizes several other international conferences, including ICAMSF 2025. His international recognition includes chairing conferences, delivering keynote talks across India, the United States and China, and receiving best research paper awards. Currently providing consultancy to the health industry for biomedical device manufacturing machines, he has filed over 30 patents with 4 international and 18+ additional patents already granted. His interdisciplinary research contributions, particularly at the intersection of the engineering, health and manufacturing sectors, have led him to edit or author five books and serve on editorial boards of numerous reputed international peer-reviewed journals.

Atul Babbar, PhD, is the Deputy Dean (Research and Development) and an Assistant Professor at Shree Guru Gobind Singh Tricentenary University (SGT), Haryana, India. He earned a PhD at the Thapar Institute of Engineering and Technology, Patiala, India, where he specialized in neurosurgical bone grinding—a procedure removing bone sections to create passages for accessing tumors in the brain and spine. His research addressed critical concerns of neurosurgeons regarding real-time temperature monitoring, thermal tissue injury and depth of thermogenesis and osteonecrosis. Dr. Babbar's research interests span interdisciplinary fields where engineering and medical research intersect, including medical devices, manufacturing, hybrid machining processes and biomedical and additive manufacturing. He has written more than 60 research articles in various international and national web of science and Scopus-indexed journals. His intellectual property portfolio includes 20 published Indian patents and 4 granted international patents. Dr. Babbar serves as a guest editor for various reputed journals and as a book series editor. He has reviewed numerous research articles for peer-reviewed journals and published multiple international books and book chapters. Dr. Babbar is an Associate Member of the Institution of Engineers (AMIE), India, and a Professional Member of the Association for Computing Machinery.

Contributors

Ajay Kumar Badhan earned a BTech in Computer Science Engineering from Jawaharlal Nehru Technological University, Kakinada, India, in 2011, and an MTech in Computer Science and Technology from Gitam University, Visakhapatnam, India, in 2014. He is a PhD scholar in Computer Science and Engineering at Lovely Professional University, India. Certified in National Programme on Technology Enhanced Learning (NPTEL) courses such as programming in Java with a gold medal, Internet of Things with a silver medal and software engineering with elite, he has received a merit certification on the Internet of Things (an online contest by Texas Instruments University Program conducted in June 2019). Badhan is an active member of the International Association of Engineers (IAENG), and his primary research interests include the Internet of Things and machine learning.

Patel Amishkumar Bhagvandas is a Piping Engineering Resource Leader based in Richmond, Texas, USA. With expertise in piping design, project management and engineering solutions, he leads critical projects in the energy and industrial sectors. His technical proficiency and leadership ensure efficient pipeline systems, compliance with industry standards and optimized resource allocation for complex engineering challenges.

Abhishek Bhattacherjee, PhD, is Associate Professor in the Department of Computer Science Engineering at Lovely Professional University. He earned a PhD in Computer Science Engineering from Chitkara University and has over 15 years of academic experience. His research interests include cloud computing, artificial intelligence, machine learning, cybersecurity and software engineering. Throughout his career, Dr. Bhattacherjee has contributed significantly to academia through numerous publications in reputed journals and international conferences. He has actively guided student research projects across various domains, including cloud microservices, AI-driven applications, cybersecurity frameworks and relevant technology.

Anupam Bonkra is Assistant Professor in the Department of Computer Science and Engineering at Maharishi Markandeshwar (Deemed to be University), Haryana, India. He has expertise in the areas of machine leaning and deep leaning. He is an experienced academician and researcher who has written several noteworthy publications.

Muralimohan Cheepu, PhD, is Research Manager in Welding Processes and Wire Arc Additive Manufacturing Developments at Vitzronextech Co., Ltd, Welding Research Institute, South Korea. Earlier Dr. Cheepu was a Postdoctoral Researcher in the Welding and Joining Laboratory at Pukyong National University, South Korea. He earned a BE in Mechanical and

Production Engineering from Andhra University, Visakhapatnam, India; an MTech in Welding Engineering from the National Institute of Technology (NIT) Tiruchirappalli, India; and a PhD in Manufacturing from the Kyungsung University, Busan, South Korea. Dr. Cheepu serves as a reviewer for various manufacturing and materials-related journals, and he is an editorial board member of the *Journal of Metallic Material Research* and the *Journal of Material Science and Technology Research*. Dr. Cheepu has written one book on additive manufacturing, and has filed two patents. He worked as a Junior Manager at JSW Steel Ltd, Bellary, India, to fabricate boilers and maintain heavy equipment in sinter and blast furnace units.

Sriharsha Chollangi is Senior Systems Engineer at Altice, Long Island City, New York, USA. He was formerly a System Administrator at Boston Scientific and Optimum. Chollangi earned a BTech in Electrical and Electronics Engineering from Jawaharlal Nehru Technological University, Hyderabad, India, and an MTech in Electrical and Electronics Engineering from Gannon University, Pennsylvania, USA.

Neeta Deshpande, PhD, is Associate Professor in the Department of Management Studies and Research, Gokhale Education Society's R. H. Sapat College of Engineering, Nashik. She has 26 years of teaching experience, and her research interests include domain data mining, AI and multimedia. Dr. Deshpande has written 56 articles for various national and international journals.

Saurav Goel, PhD, is Professor of Manufacturing at London South Bank University (LSBU), UK. He has published more than 200 articles with >6500 citations (Google Scholar), and he has been ranked among the Top 2% scientists of the world for the past five consecutive years.

Rahul Grover, PhD, is Assistant Professor in the Department of Civil Engineering, Maharishi Markandeshwar University, Mullana, Ambala Haryana, India. His expertise includes construction and structure engineering. He has published several journals and conferences articles.

Shaman Gupta, PhD, is Assistant Professor in the Department of Mechanical Engineering at Maharishi Markandeshwar University, Ambala, Haryana, India. He earned a BTech and an MTech in Mechanical Engineering from Kurukshetra University, Haryana, India, and a PhD from the National Institute of Technology (NIT) Kurukshetra, Haryana, India. Dr. Gupta's main research area is industrial engineering.

Md Umar Ibrahim is a master's student in robotics and AI scholar with a BTech in Mechanical Engineering earned from Jamia Millia Islamia, India, in 2024. His research interests lie in robotics and autonomous systems, and he focuses on intelligent, adaptive machines and their real-world applications.

Vijaykumar S. Jatti, PhD, is Professor in the Department of Mechanical Engineering, School of Engineering and Applied Sciences, Bennett University, Noida, India. He earned a BE in Mechanical Engineering from Visvesvaraya Technological University (VTU), Belgaum, India; an MTech in Mechanical Engineering from the National Institute of Technology (NIT) Calicut, India; and a PhD in Mechanical Engineering from Symbiosis International (Deemed University), Pune, India. Dr. Jatti has more than 13 years of teaching, research and industrial experience. He has published 106 papers (WoS-7, Scopus-67, other indexed-10, IC-17 and NC-5) in reputed national and international journals and national and international conferences. Dr. Jatti has published 10 patents and written 12 book chapters and 11 books with national and international publishers. He is also a reviewer for many international journals and conferences. Dr. Jatti has completed one minor research project of 1.5 lacs funded by SIU, Pune, and served as a co-guide for three PhD research scholars and two PhD research scholars. He has guided 24 postgraduate projects and 14 undergraduate projects. Dr. Jatti has given 46 invited talks at different engineering colleges, universities and industrial events, and he has received 16 awards and recognitions from different organizations, institutions and agencies for his contribution to academics and research work.

Rupinder Kaur is presently working in the Department of Information Technology, Guru Nanak Dev Engineering College, Punjab, India. She has vast academic and research experience in the domains of digital watermarking of images, machine learning and energy optimization.

Hreetabh Kishore, PhD, earned a PhD in Mechanical Engineering from the Indian Institute of Technology Ropar, Rupnagar, Punjab, India, in 2023. He is Instructor in the Department of Manufacturing Engineering Technology, Tennessee Tech University, Cookeville, Tennessee, USA. Dr. Kishore's research areas include precision micro-manufacturing, additive manufacturing, material modeling, optimization and mechanical processes simulation.

Abhinav Kumar, PhD, is a distinguished researcher and Senior Scientific Officer at the Department of Nuclear and Renewable Energy, Ural Federal University (UrFU), Russia. With a strong academic and research background, he has made significant contributions to the fields of renewable energy systems and sustainable technologies. Dr. Kumar has worked in energy systems research for the last 9 years on solar PV energy, parabolic trough solar collectors, advanced solar cell materials, water-splitting applications, battery storage, superconducting magnetic energy systems, etc. His computational skills include Comsol and Ansys. Dr. Kumar has published high-quality articles and presented his research at international forums. He collaborates with national and international research institutions, fostering the exchange of knowledge in energy science. Dr. Kumar's work aligns with global efforts to address climate change, energy security and decarbonization strategies.

Raman Kumar, completed his BTech (Mechanical Engineering) and MTech (Production Engineering) from Guru Nanak Dev Engineering College, Ludhiana (Punjab), and his doctorate (Mechanical Engineering) from I. K. G. Punjab Technical University, Jalandhar—Kapurthala, India. With 5 years of industry experience and 19 years in teaching and research, his expertise spans advanced manufacturing technologies, sustainable machining and optimization techniques. His work integrates machine learning (ML) and artificial intelligence (AI) to enhance machining efficiency, material properties and process optimization in both industrial and biomedical applications. Aligned with the mission of sustainable and environmentally friendly technologies, Dr. Kumar's research focuses on additive manufacturing, composite materials and nanotechnology, where experimental modeling and optimization techniques such as multi-criteria decision-making, response surface methodology, ML, AI and bibliometric analysis play crucial roles. His expertise includes renewable energy systems, particularly improving energy efficiency and sustainability in smart manufacturing and industrial contexts, including Industry 4.0 and the emerging Industry 5.0. As an active contributor to the academic and industrial communities, Dr. Kumar has to his credit over 130 research publications. His work has been recognized globally, and he is included in Stanford University's Top 2% Scientists for 2022, 2023 and 2024. Dr. Kumar also serves as an Associate Editor for several Frontiers journals, covering psychology, public health, climate and rehabilitation sciences. In addition, he reviews for leading journals such as the *Journal of Cleaner Production*, *Sustainable Cities and Society* and *IEEE Sensors Journal*. Dr. Kumar is a life member of the International Society for Technology in Education (ISTE) and SAE India, and he was awarded the 2021 ISTE Best Teacher Award (Section) and the 2023 Faculty Excellence Award by the GNDEC Alumni Association. He also received the Best Paper Award at an international conference.

K. P. Lijesh, PhD, is Postdoctoral Researcher at Louisiana State University (LSU), Baton Rouge, Louisiana, USA. He earned a PhD in Mechanical Engineering from the Indian Institute of Technology (IIT) Delhi, with expertise in entropy generation, magnetic bearings, journal bearings and MR fluids. His research focuses on improving mechanical system reliability through tribology and thermal analysis. Dr. Lijesh has published widely in reputed journals and actively contributes to the tribology research community.

Wulfran Fendzi Mbasso, PhD, earned a BTech (2014) in Electrical and Industrial Computing Engineering from the University of Doualal, Cameroon; an MSc (2016) in Electrical Engineering from the University of Ngaoundéré, Cameroon; and an MRes (2020) in Electrical Engineering and a PhD (2024) in Electrical Engineering from the University of Douala, Cameroon. His research interests include power system control, optimization techniques, automated systems and electronics. Dr. Mbasso has a robust background in electrical engineering and industrial computing engineering, particularly focusing on optimization of hybrid renewable energy systems, and he is a reviewer for Elsevier journals.

Omar Mbrouk, PhD, is Researcher in Environmental Science with a specialization in chemical science at the University of Sadat City (USC) in Sadat City, Egypt. His research focuses on water treatment, green hydrogen production and plasmonic composite nanomaterials. Dr. Mbrouk is also associated with nano-photochemistry and its environmental applications at the Environmental Studies and Research Institute (ESRI) within USC. He has published several articles in international peer-reviewed journals.

Raghvendra Kumar Mishra, PhD, EPSRC (UK RI), is a distinguished academic and researcher specializing in nanomaterials, polymer composites and their diverse applications. He earned a PhD from Cranfield University, UK, funded by the Engineering and Physical Sciences Research Council (EPSRC) from 2019 to 2022, where he delved into the intricacies of nanomaterials and polymer composites within the School of Aerospace, Transport and Manufacturing. Dr. Mishra is Materials Scientist in the Chemical Engineering Department at the Indian Institute of Technology (IIT) Delhi. He was honored as the 2019 Excellent Researcher of the Year (Male) by GERA and received the prestigious Fellowship of the Engineering and Physical Sciences Research Council, UK (EPSRC). His research articles, conference papers and invited talks showcase his dedication to advancing scientific knowledge in his field. Dr. Mishra's impactful contributions to the field of nanomaterials and polymer composites, his extensive research experience and his dedication to academia make him a prominent figure in the scientific community.

Ujval Mutyala is an active researcher of quantum computing, cybersecurity and cloud computing, exploring the future implications of emerging technologies on digital security. He earned a master's degree from Eastern Illinois University, Charleston, Illinois, USA, and he has more than a decade of extensive experience in the field of cybersecurity. Mutyala has contributed to securing critical infrastructure and enterprise systems for multiple organizations across a diverse range of industries and geographies. His expertise spans threat detection and response, risk assessment, vulnerability management and the implementation of robust security architectures. Mutyala is an active contributor to network security, cloud security and identity and access management, ensuring compliance with industry standards and regulatory frameworks. He currently holds memberships with ISC^2, IEEE, ACM and BCS.

S. M. Muzakkir, PhD, is Professor in the Department of Mechanical Engineering, Faculty of Engineering and Technology, Jamia Millia Islamia, New Delhi, India. He earned a BSc in Mechanical Engineering (1990) from Jamia Millia Islamia and an MTech in Design of Mechanical Equipment (1996) and a PhD in Tribology (2015) from the Indian Institute of Technology (IIT) Delhi. With 34 years of teaching experience, his research interests include tribology (including nanolubricants, wear, journal bearings, magnetic bearings and MR brakes), mechanical vibrations (encompassing modal analysis and fault diagnostics) and innovative product design. Dr. Muzakkir has written 45 Scopus-indexed and 39 non-Scopus research papers, presented 21 international conference papers and contributed five book chapters. He also holds four granted patents and two published patents, and he has received five distinguished awards. Dr. Muzakkir has supervised 18 MTech theses and one PhD thesis. His dedication and excellence have been recognized with five distinguished awards.

Suresh Pratap, PhD, is Assistant Professor in the Department of Mechanical Engineering at IIMT College of Engineering, Greater Noida, Uttar Pradesh, India. He earned a PhD in Engineering from Birla Institute of Technology, Mesra, where he also worked as a Research Fellow and Teaching Assistant. His doctoral research focused on reverse micro electrical discharge machining. Dr. Pratap has expertise in microfabrication, advanced manufacturing processes and integration of artificial intelligence in manufacturing processes. He has published more than 10 SCIE-indexed papers and holds a patent in data science workflow. Dr. Pratap has actively participated in national and international conferences, and he is a reviewer for multiple journals. His skills include MATLAB, ANSYS and advanced manufacturing tools such as CNC, μEDM and 3D printing. Dr. Pratap is also involved in various All India Council for Technical Education faculty development programs, and he is a member of professional bodies such as the ASME and ASTM.

Aseeb Sabu is a Project Trainee at Tata Consultancy Services (TCS), and he earned a BTech in Mechanical Engineering from Jamia Millia Islamia (2024). He is BSc in Data Science scholar at the Indian Institute of Technology (IIT) Madras. His research interests include robotics, autonomous systems and smart mechanical design. Sabu has actively contributed to interdisciplinary projects and presented research at international conferences.

A. Saiyathibrahim, PhD, earned a PhD in Mechanical Engineering with an emphasis on functionally graded materials from Anna University, Chennai, India. His research interests lie in the broad areas of advanced machining, centrifugal casting, functionally graded materials and additive manufacturing. Dr. Saiyathibrahim has published several book chapters, scientific journals and a patent. In addition to his 13 years of teaching experience, he has been associated with professional societies such as the ISME, IAAM, IAAC and ISDS.

Ankit Sharma, PhD, is Associate Director (Research) at the Centre for Research Impact and Outcome (CRIO) Department at Chitkara University, Punjab, India. He earned a PhD in Mechanical Engineering from Thapar Institute of Engineering and Technology, India. He has more than 10 years of experience in academics, research, consulting, training and industry. Dr. Sharma also organized several international conferences, including ICEMSMCI 2023, ICEMSMCI 2024 and ICAMSF 2025, and he served as the general chair and convenor. He has authored numerous national and international publications in SCI, Scopus and Web of Science-indexed journals. Dr. Sharma has given several invited seminars and keynote talks at international (US, China and India) conferences and seminars, and he has received several best research paper awards. He is also the reviewer board member of several international journals published by Elsevier, Sage, Springer, Taylor & Francis, MDPI, etc.

Pallavi Sharma, PhD, is a machine learning expert and Adjunct Faculty at the Department of Mechanical Engineering, Islamia University of Bahawalpur, Iraq. She specializes in artificial intelligence and machine learning applications for mechanical systems, predictive maintenance and thermal engineering optimization. Dr. Sharma's work integrates neural networks, deep learning and data-driven modeling to enhance industrial automation and energy efficiency. As an educator and researcher, she bridges AI with mechanical engineering, contributing to smart manufacturing and sustainable engineering solutions. Recognized for her interdisciplinary collaborations, Dr. Sharma advances intelligent systems for industrial and environmental applications, inspiring innovation in AI-driven engineering.

Sergey E. Shcheklein, DrSci, is a distinguished Russian scientist and professor specializing in thermal physics, nuclear energy, and hydrogen energy technologies. He is the Head of the Department of Nuclear Power Plants and Renewable Energy Sources at Ural Federal University (UrFU) in Yekaterinburg, Russia. Professor Shcheklein graduated from Ural Polytechnic Institute (now UrFU) and has since dedicated his career to research and education in energy-related fields. His scientific work focuses on thermal hydraulics, safety of nuclear reactors and the development of hydrogen energy systems. He has written numerous publications in leading journals, contributing significantly to advancements in energy technology. Professor Shcheklein is an active researcher with multiple citations reflecting his impact in the field. He supervises graduate students and participates in international collaborations, promoting innovative energy solutions. At UrFU, he plays a key role in academic leadership, fostering research in renewable energy and nuclear engineering.

Shyan is intern at the Advanced Research and Development Center, LIPS Research Foundation, which is affiliated with the European International University, France. She is working on machine learning models.

Balwinder Singh is Assistant Professor in the Electrical Engineering Department, Guru Nanak Dev Engineering College (GNDEC), Ludhiana, Punjab. He earned a BTech and an MTech from GNDEC. He has more than

12 years of teaching experience. His areas of interest include power systems, electrical machines and renewable energy sources.

Harvinder Singh, PhD, is Associate Professor in the Department of Mechanical Engineering, Chandigarh Engineering College Landran. He earned a PhD from Chandigarh University, an MTech from Sant Longowal Institute of Engineering and Technology, Punjab, and a BTech from Guru Jambheshwar University of Science and Technology, Hisar. He has more than 13 years of teaching experience. Dr. Singh has published more than 50 papers in reputed national and international journals and conferences.

Kunwar Partap Singh is Assistant Professor in the Department of Electronics and Communication Engineering at Guru Nanak Dev Engineering College, Ludhiana, Punjab, India. He earned a BTech in Electronics and Communication Engineering from Guru Nanak Dev University (GNDU), Amritsar, in 2016, and an MTech in Electronics and Communication Engineering at GNDU in 2018. His primary areas of interest include nanoelectronics, with a specific focus on nanowire fabrication and the use of DNA-based structures for nanowire development. He has actively engaged in research that explores the intersection of biology and electronics, aiming to contribute to the advancement of molecular-scale electronic components. With a passion for both teaching and research, he continues to mentor students and contribute to emerging technologies in the field of nanoelectronics.

Sehijpal Singh, PhD, is Professor of Mechanical and Production Engineering at Guru Nanak Dev Engineering College, Ludhiana, Punjab, India. He has made significant contributions to the field of nontraditional machining processes, decision-making and optimization techniques, and his research has garnered over 5,900 citations. Dr. Singh's h-index stands at 37, while his i10-index is 66. Among his notable works is the study on the parametric optimization of powder mixed electrical discharge machining using response surface methodology, which has been widely cited and serves as a foundational reference in the domain.

Sukhpal Singh is Assistant Professor in the Department of Electrical Engineering, Guru Nanak Dev Engineering College, Ludhiana, Punjab, India. He earned a BTech in Electrical Engineering and an ME in Power Systems and Electric Drives, and he is currently a PhD scholar in electrical engineering. He has six years in industry and 12 years of teaching and research experience. Singh's areas of interest include power generation, transmission and distribution; smart grids and microgrids; renewable energy integration (solar and wind); signal processing; neural engineering; sustainable energy solutions; demand response optimization; and power system stability. He has taught power systems, electrical machines, energy efficient machines, power electronics, signals and systems, circuit analysis, smart grids, flexible AC transmissions, high-voltage engineering, etc., at the undergraduate and postgraduate levels. Singh has more than 20 research publications in national and international conferences and reputed journals. He is a reviewer for *Frontiers in Psychology*, and he is a Life Member of the International Society for Technology in Education (ISTE). Singh has consulted for various industrial projects in designing sewage treatment plants and electrical power stations, and he has carried out energy audits in various industries and suggested solutions for problems. Academically, he has participated in various faculty development programs, conferences and expert lectures.

Swarn Singh is Assistant Professor in the Department of Mechanical Engineering, Chandigarh Engineering College, Landran, India. He earned a BTech and an MTech from I. K. Gujral Punjab Technical University, Jalandhar. He has more than 16 years of teaching experience. He is also a PhD scholar at Maharaja Agrasen University, Solan, India. He has published more than 15 research papers and book chapters in various reputed journals. Singh has attended and presented papers at many national and international conferences.

Priyanka Sinha, PhD, earned a master's degree in Library Science from Delhi University, India, and a PhD in Library Science from Panjab University, Chandigarh, India. She is Assistant Librarian at Lamrin Tech Skills University, Punjab, with more than three years of experience as a library professional. Her research interests include digital literacy, research data management, artificial intelligence, applications of the Internet of Things, bibliometrix visualization and dissemination.

Jai Sonker, PhD, is Senior Professor in the Amity School of Hospitality at Amity University, Rajasthan, India. He is also an editorial board member of *International Journal of Hospitality* and tourism studies. Dr. Sonker is a lifetime member of the All India Council for Technical Skill Development. He has received Best Academic Coordinator of Global Education and Corporate Leadership Awards. Dr. Sonker's research and administrative contributions are significant recent technological advancements.

Pushpendra Ratnam Verma, PhD, is a seasoned academician and researcher with over a decade of distinguished experience in computer science and engineering. He is Assistant Professor at Lovely Professional University, Punjab, India, and he previously taught at Himgiri Zee University, Uttarakhand, India. A PhD scholar in underwater sensor localization, Dr. Verma's scholarly contributions span international patents, IEEE conferences and reputed journals. His expertise lies in artificial intelligence, machine learning, wireless sensor networks, optimization algorithms and secure communication protocols. With a strong commitment to academic excellence, he has guided postgraduate dissertations, contributed to curriculum development and played key roles in academic administration and event organization. A recipient of prestigious accolades, including the Best Faculty of the Year and NAAC Excellence awards, Dr. Verma is a proactive member of the IEEE and the Sensors Council. He epitomizes a dynamic blend of innovation, pedagogy and research-driven leadership in modern engineering education.

Padmini Yenumula is Research Scholar in the Department of Food Science and Technology, Pukyong National University, Busan, Republic of Korea. She is a passionate researcher working toward the development of cutting-edge technologies.

1

Advanced Manufacturing Techniques of Additive Manufacturing Processes: A State-of-the-Art Review and Future Applications

Abhishek Bhattacherjee, Ajay Kumar Badhan, Pushpendra Ratnam Verma, Raman Kumar, Rupinder Kaur, Sehijpal Singh, and Wulfran Fendzi Mbasso

1.1 Introduction

The production landscape in the present process is a paradigm shift with the appearance of emerging technologies that bridge the space between virtual and physical structures. Among these, additive manufacturing (AM), popularly known as three-dimensional (3D) printing, has emerged as a disruptive force, presenting remarkable flexibility, customization, and design freedom in producing parts and additives. Not like conventional subtractive production techniques that eliminate material from a strong block, AM builds gadgets layer by layer immediately from virtual models, significantly decreasing material waste and manufacturing time. Concurrently, the continuing Industry 4.0 revolution, characterized by combining cyber-physical structures, the Internet of Things (IoT), cloud computing, and artificial intelligence (AI), is reshaping how manufacturing structures perform, screen, and optimize themselves. While fused with AM, those technologies release effective new capabilities, remodeling additive approaches into smart, related, and self-reliant manufacturing ecosystems (Bendoly et al., 2024). This chapter explores the mixing of AI and Industry 4.0 technology with AM, presenting insights into how this convergence complements technique reliability, quality of product, and operational efficiency. It delves into core principles, technological enablers, real-world applications, and destiny developments shaping this alteration. The subsequent segment offers a top-level view of AM, putting the inspiration for know-how on how AI and enterprise 4.0 technologies are being embedded into its methods (Budzik et al., 2022).

DOI: 10.1201/9781003506492-1

1.1.1 Overview of AM Processes

Additive production (AM) refers to technologies that create 3D items by sequentially depositing material layers, primarily based on virtual layout data. It enables the fabrication of complex geometries that are hard or impossible to achieve through traditional production techniques. AM has transitioned from a prototyping device to a viable alternative for quick-use component production in industries with aerospace, automobile, healthcare, and customer electronics (Chen et al., 2022).

Numerous AM technologies are shown in Figure 1.1, each using special substances and strategies. The most commonly used technologies encompass:

i. **Fused Deposition Modeling (FDM):** Melts thermoplastic filament and deposits fabric layer through layer.
ii. **Stereolithography (SLA):** Makes use of ultraviolet (UV) light to solidify liquid resin into hardened plastic.
iii. **Selective Laser Sintering (SLS) and Selective Laser Melting (SLM):** Use laser power to fuse powdered substances.
iv. **Electron Beam Melting (EBM)** – Employs an electron beam for high-precision steel part fabrication (Dehghan et al., 2025).

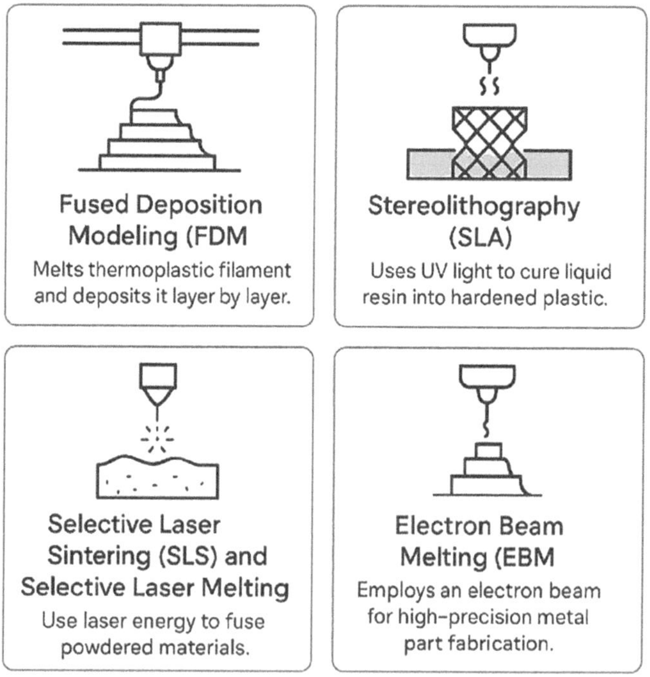

FIGURE 1.1
Types of AM technologies.

The primary advantages of AM include:

- Layout flexibility – Capacity to manufacture complicated and customized designs.
- Material efficiency – Minimal waste as compared to subtractive techniques.
- Rapid prototyping – Shorter lead instances from idea to prototype.
- Decentralized manufacturing – On-demand production in the direction of point of use.

Despite these advantages, AM faces demanding situations together with constrained material alternatives, inconsistent excellencies, slow construct speeds, and high manufacturing expenses for certain programs (Gao et al., 2015). These challenges open the door for superior technologies together with AI and enterprise 4.0 components to optimize and automate AM workflows, decorate reliability, and allow mass customization at scale. In subsequent sections, we will discover how enterprise 4.0 and AI are reshaping the AM landscape via introducing new potential across the whole cost chain, from layout and procedure control to quality warranty and supply chain integration (Gupta et al., 2021).

1.1.2 Rise of Industry 4.0 and AI in Manufacturing

The fourth revolution, extensively known as enterprise 4.0, represents a transformative technology in manufacturing characterized by the convergence of advanced digital technologies with traditional manufacturing structures, not like preceding industrial revolutions that had been pushed by way of mechanization, electrification, or automation. Industry 4.0 is defined by a combination of cyber-physical structures (CPS), IoT), cloud computing, and AI into each factor of producing operations. The modification allows factories to become intelligent, interconnected, data-driven, promoting real-time transparency, predictive capabilities, and self-directed decision-making. Machines and systems at the moment are prepared with sensors and connectivity, taking into account non-stop monitoring of operations, dynamic optimization of sources, and seamless human–machine collaboration (Hendrixson, 2017). These clever environments are capable of self-analysis, self-optimization, and even self-restoration, making production more efficient, flexible, and resilient. Among all allowing technologies, AI has emerged as the relevant intelligence layer that powers enterprise 4.0. Algorithms for device learning, computer vision, and deep learning models are being implemented to analyze large volumes of data generated by manufacturing systems. These AI techniques are enabling:

a. Predictive maintenance by forecasting equipment failures before they occur.

b. Technique optimization through adaptive mastering and manage structures.
c. Fine assurance through actual-time illness detection and anomaly identity.
d. Supply chain intelligence through demand forecasting and inventory optimization.

On this context, the synergy between AI and Industry 4.0 is riding a profound shift from rule-based totally automation to self-sustaining and AM structures (Hernandez Korner et al., 2020). This evolution is particularly impactful for additive production (AM), wherein variability, procedure complexity, and fabric behaviors require smart interventions for constant and consistent outputs. The software of AI in AM, inside the Industry 4.0 framework, enables smarter design through generative strategies, real-time tracking of build strategies, automated post-processing, and virtual twins for predictive simulation. As such, the rise of Industry 4.0 and AI is not merely an enhancement of current procedures; it is redefining how merchandise is conceived, produced, and delivered. The following section delves into the foundational additives of Industry 4.0 and explores how those digital technologies offer the backbone for integrating intelligence into AM environments (Khorasani et al., 2022).

1.1.3 Importance and Scope of Integrating AI and Industry 4.0 with AM

Combining AI and enterprise 4.0 technologies with AM reshapes modern-day production paradigms by allowing smart, records-driven, and agile production systems. While AM already offers great advantages consisting of layout flexibility, reduced material waste, and speedy prototyping, its full ability stays underutilized due to demanding situations like inconsistent high-quality, restrained scalability, and complicated technique management (Khajavi et al., 2014). The fusion of AI and Industry 4.0 technologies addresses those barriers and paves the way for a brand-new era of smart additive manufacturing ecosystems.

1.1.3.1 Importance of Integration

- More suitable process optimization

  AI algorithms can examine procedure parameters, cloth behaviors, and environmental situations to suggest or routinely practice safe settings. This results in advanced print quality, decreased defects, and lower electricity intake, as shown in Figure 1.2.

- Real-time monitoring and management

  Sensors, IoT devices, and device imagination and perception systems, including AI, allow real-time tracking of layer formation, temperature

FIGURE 1.2
Benefits of AI integration.

profiles, and construct consistency. These systems allow for dynamic adjustments, preventing errors throughout the build procedure (Kumar et al., 2019).

- Predictive upkeep and downtime discount

  AI systems trained on historical and real-time device statistics can anticipate factor disasters before they occur. This minimizes unplanned downtime, extends system lifespans, and ensures nonstop operation.

- Mass customization and agile production

  Through combining AI's potential to generate optimized designs (Lee et al., 2014).

- Advanced supply chain performance

  Industry 4.0 connectivity supports decentralized and disbursed manufacturing models. Elements can be published on a website or directed at the end user, reducing lead instances, logistics expenses, and carbon footprints.

- Sustainability and waste discount

  Smart management structures and AI-driven simulations allow for material savings, strength performance, and better lifecycle control, assisting sustainable manufacturing practices.

1.1.3.2 Scope of Integration

The scope of integrating AI and Industry four.0 with AM spans the whole product lifecycle, along with:

- Design and preprocessing: AI-powered computer-aided design (CAD) tools and topology optimization for lightweight, purposeful designs.
- In-manner tracking: Use of AI to gain knowledge of real-time anomaly detection and adaptive process manipulation.
- Put up-processing: Automatic support removal, first-rate inspection using computer vision, and feedback loops to improve destiny builds (Li et al., 2021).
- Quality assurance: Records-pushed illness prediction and certification of printed elements via digital information.
- Employer-level integration: Seamless data goes with the flow across enterprise structures (e.g., Enterprise Resource Planning (ERP), Manufacturing Execution System (MES), and Product Lifecycle Management (PLM)), permitting complete traceability and compliance.

Incorporating all the above also helps move disciplinary convergence, connecting substance science, mechanical engineering, records technological know-how, and cloud infrastructure to create holistic production solutions (Lu, 2017). Combining AI and enterprise 4.0 technology with AM is not an incremental improvement; it represents a strategic enabler for next-generation production, facilitating smarter, more flexible, and more sustainable production structures. In the next sections of this chapter we will explore the technological foundations and real global programs of this integration in more detail, as shown in Figure 1.3.

1.2 Fundamentals of AM

AM, usually known as 3D printing, is a manufacturing process that builds objects layer by layer from digital models. In contrast to conventional subtractive manufacturing, which removes material from a stable block, AM creates additives by including material only in the desired location, thereby presenting massive advantages in layout freedom, metal utilization, and customization (Mourtzis et al., 2018).

1.2.1 Principles and Types of AM

At the core of all AM tactics lies a shared principle: material is deposited, joined, or solidified layer by layer to shape a 3D item based on a virtual design, usually inside the shape of a CAD document. The process generally involves the following steps:

- Design: Advent of a 3D CAD version.
- Reducing: Conversion of the model into thin horizontal layers and the technology of toolpaths.
- Material deposition or solidification: Layer-by-layer fabrication using a particular AM system.
- Later processing: Cleaning, curing, or floor finishing for final utility.

The ISO/ASTM 52900 standard classifies AM into several classes based on the approach of material merging or deposition among them. The following are the most widely used:

a. FDM

Principle: Thermoplastic filament is heated and extruded via a nozzle, depositing fabric layer through layer on a construct platform (Sani et al., 2024).

Materials: Polylactic acid (PLA), acrylonitrile butadiene styrene (ABS), polyethylene terephthalate glycol (PETG), thermoplastic Polyurethane (TPU), nylon.

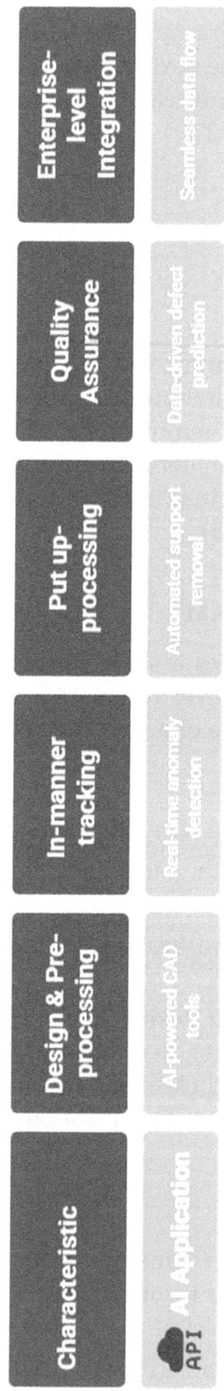

FIGURE 1.3
AI and Industry 4.0 integration with AM.

Packages: Prototyping, tooling, low-price practical parts.

Pros: Value-powerful, smooth to use, broadly accessible.

Cons: Lower resolution and electricity as compared to different AM sorts.

b. SLA

Principle: Makes use of a UV laser to cure liquid photopolymer resin layer by means of layer.

Materials: Photosensitive resins with numerous mechanical houses.

Packages: Dental models, jewelry, precision prototypes (Shafiee, 2025).

Pros: Excessive resolution, wonderful surface end.

Cons: Brittle materials, post-curing required, touchy to UV degradation.

c. SLS

Principle: Makes use of a laser to selectively fuse powdered thermoplastic material (generally nylon) layer by means of layer (Singh et al., 2021).

Materials: Nylon, TPU, composites.

Packages: Practical prototypes, mechanical components, custom enclosures.

Pros: No support structures wished, suitable mechanical properties.

Cons: High device fee, rough floor finish.

d. Direct metal laser sintering (DMLS)/SLM

Principle: A high-powered laser fuses or melts metal powder to shape strong metallic parts.

Material: Chrome steel, titanium, aluminum, cobalt-chrome, Inconel.

Packages: Aerospace, medical implants, complex steel additives.

Pros: Sturdy, dense elements; complicated geometries are manageable.

Cons: High-priced equipment and materials, requires advanced publishing (Tang et al., 2016).

These technologies represent the range of AM processes regarding substances, complexity, precision, and business relevance. Selecting a suitable AM approach relies upon the software requirements, fabric properties, floor end, element geometry, and price constraints. As we delve similarly into the combination of AI and Industry 4.0 with AM, it is crucial to comprehend how the precise traits of each AM manner can be enhanced, monitored, and optimized using intelligent technology, as shown in Figure 1.4.

1.2.2 Materials Used in AM

Material selection is critical in determining the mechanical properties, look, functionality, and alertness of components produced via AM (Thompson et al.,

Characteristic	FDM	SLA	SLS	DMLS/SLM
Principle	Thermoplastic filament extrusion.	UV laser cures photopolymer.	Laser fuses thermoplastic powder.	Laser fuses/melts metal powder.
Materials	PLA, ABS, PETG, TPU, Nylon	Photosensitive resins	Nylon, TPU, composites	Stainless steel, titanium, aluminum
Applications	Prototyping, tooling, parts	Dental models, jewelry, prototypes	Prototypes, components, enclosures	Aerospace, implants, additives
Pros	Cost-effective, easy to use	High resolution, surface finish	No supports, good mechanics	Strong, dense, complex geometries
Cons	Lower resolution, strength	Brittle, post-curing, UV sensitive	High cost, rough finish	Expensive, advanced post-processing

FIGURE 1.4
Additive manufacturing technologies comparison.

2016). Over the years, fabric technological know-how improvements have notably broadened the scope of AM, enabling the production of prototypes and purposeful products across various industries. AM substances are generally decided on the basis of the preferred attributes which include energy, flexibility, heat resistance, biocompatibility, and value. The substances used in AM may be categorized into the following most essential classes as shown in Figure 1.5:

- Thermoplastics: Thermoplastics are some of the most commonly used substances in AM, especially in FDM.
- PLA: Biodegradable, clean to print, ideal for prototyping and academic uses.
- ABS: Long-lasting and impact-resistant, used for functional components.
- PETG: Combining energy and flexibility with suitable chemical resistance.
- Nylon: Known for high power and versatility, used in industrial and mechanical applications.
- TPU: Flexible, rubber-like fabric for gentle elements.
  - Photopolymers: Photopolymers are used on the whole in SLA and digital light processing, and they are exposed to UV light.
- Standard resin: High element and smooth surface, suitable for models and prototypes.
- Engineering resin: Tailor-made for toughness, warmth resistance, or flexibility.
- Dental/biocompatible resin: Used for clinical programs and temporary dental systems.

Note: Photopolymers usually require post-curing and may be brittle.

- Metal Powders: DMLS and SLM strategies depend on the best metal powders to provide high-energy, high-strength components.
  - Titanium alloys (e.g., Ti6Al4V) are lightweight, strong, and corrosion-resistant, used in aerospace and biomedical implants.
  - Aluminum alloys are lightweight and have desirable thermal conductivity; they are best for cars and aerospace.
  - Chrome steel: Sturdy and corrosion-resistant, normally used for useful prototypes and tooling.
  - Cobalt-chrome: Excessive put-on resistance and biocompatibility, used in scientific implants and dentistry.
  - Inconel (nickel-based alloys): High temperature resistance for turbines and heat exchangers programs.

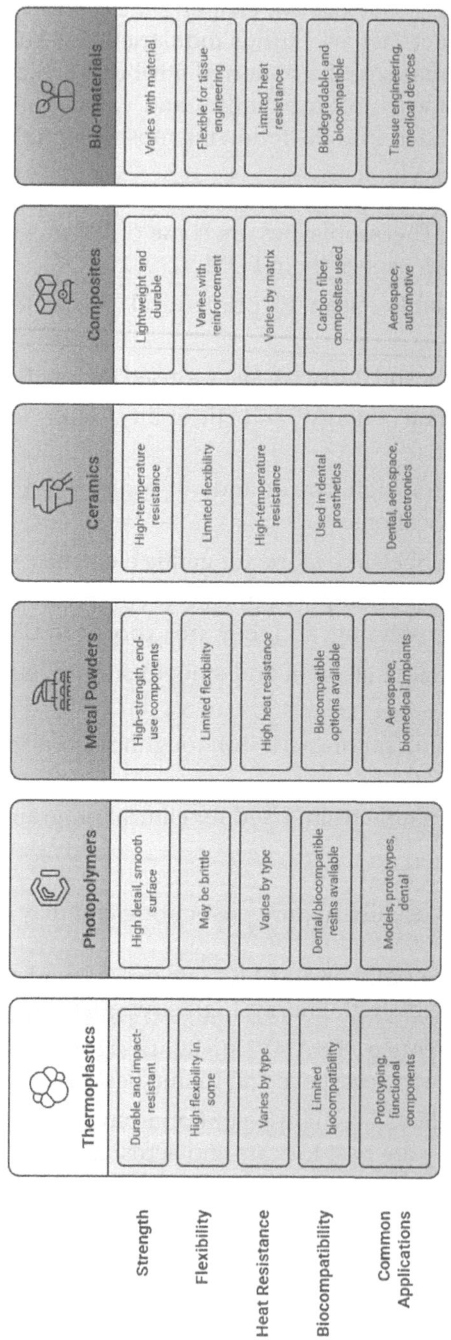

FIGURE 1.5
Additive manufacturing material comparison.

- Ceramics: Ceramic substances are used in Vat photopolymerization and Binder jetting for programs requiring excessive temperature resistance and electrical insulation.
  - Alumina (Al_2O_3)
  - Zirconia (ZrO_2)
  - Silicon carbide (SiC)

Packages encompass dental prosthetics, aerospace components, and digital parts (Ucar et al., 2024).

- Composites: Composite materials integrate a polymer or metal matrix with reinforcing fibers or debris to enhance mechanical performance.
  - Carbon fiber-reinforced polymers: Lightweight and sturdy, used in aerospace and cars.
  - Glass fiber composites: Cost-powerful reinforcement for structural applications.
- Biomaterials: These materials are often biodegradable and biocompatible and are used in bioprinting and custom clinical tool fabrication (Varela et al., 2018).
  - Hydrogels: For tissue engineering and cellular scaffolds.
  - Polycaprolactone: Biodegradable polymer used in biomedical implants.
  - Alginate and collagen: Used in gentle tissue modeling and regenerative medicine.

1.2.2.1 Material Selection Considerations

While choosing materials for AM, the following factors are taken into consideration:

- Mechanical residences: Electricity, flexibility, hardness.
- Thermal resistance: Operating temperature, heat deflection.
- Chemical resistance: Publicity to solvents or moisture.
- Biocompatibility: Used for medical or dental applications.
- Value and availability: Economic feasibility for large-scale production.

The continuous improvement of superior substances tailored especially for AM is increasing its industrial software and enabling new programs. As Industry 4.0 and AI technology evolve, fabric conduct can be better understood and managed in real time, establishing doors to material-through-design strategies and clever material choice (Wang et al., 2023).

1.2.3 Current Industrial Applications and Limitations

- Aerospace and defense

  Use: Lightweight structural components, brackets, gas nozzles, turbine blades.

  Use Case: GE Aviation's gas nozzle for jet engines, made using DMLS, is 25% lighter and five times more durable than historically manufactured parts.

  Benefits: Weight reduction, advanced fuel performance, and complicated geometries are now not viable with conventional machining, as shown in Figure 1.6.

- Car manufacturing

  Use: Custom-designed parts, tooling, rapid prototyping, and quick-use components for motorsports.

  Use Case: BMW and Ford use AM for growing prototype components, furnishings, and limited-run custom additives.

  Benefits: Reduced lead time, fee-powerful prototyping, on-demand spare element production.

- Healthcare and clinical devices

  Use: affected person-precise implants, dental prosthetics, surgical publications, bio-printed tissues.

  Use Case: 3D-published titanium implants and listening aids are customized for individual patients.

  Benefits: High personalization, stronger affected person results, reduced surgical time.

- Client goods

  Use: Eyewear frames, fashion accessories, shoe midsoles, and domestic home equipment.

  Use Case: Adidas's Futurecraft 4D shoes feature a 3D-revealed midsole optimized for performance.

  Benefits: Mass customization, innovative design, faster time-to-market.

- Architecture and construction

  Use: Complicated architectural models, 3D-published houses, and infrastructure.

  Use Case: ICON and Apis Cor have advanced 3D-published homes and shelters.

  Benefits: Value savings, reduced exertions, and environmentally friendly construction.

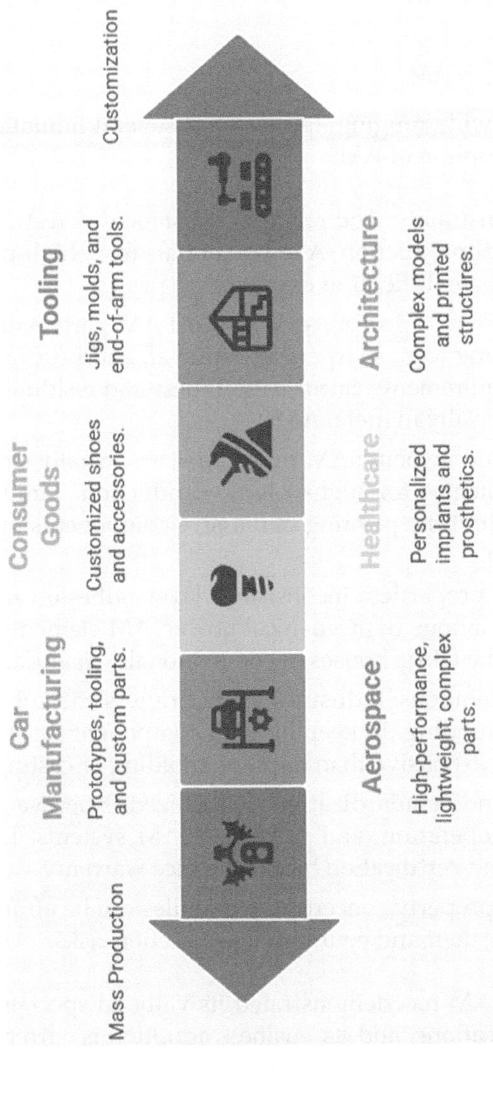

FIGURE 1.6
3D printing applications ranked by production scale and customization.

- Tooling and manufacturing aids

  Use: Jigs, furniture, molds, stop-of-arm tooling.

  Use Case: Volkswagen uses 3D-printed gear and jigs on its production lines to improve ergonomics and performance.

  Benefits: Faster new release, decreased tooling value, increased operational flexibility.

1.2.3.1 Limitations of AM

Irrespective of its wide-ranging applications, several limitations inhibit the whole business adoption of AM:

1. Material constraints: The range of substances is restricted compared to traditional production. A few materials (e.g., high-power alloys or composites) are difficult or expensive to print.
2. Surface finish and postprocessing: Most AM parts require significant postprocessing (e.g., sharpening, removal, and heat remedy) to meet business requirements. Surface roughness and residual stress troubles persist, especially in metal AM.
3. Build size and velocity: AM procedures are usually slower than subtractive strategies for high-volume production. Limitations in construct size limit the printing of massive components in an individual consultation.
4. Mechanical properties: Inconsistent layer adhesion or porosity can result in variations in mechanical power. AM elements may not usually match the fabric houses of conventionally manufactured parts.
5. Price considerations: Industrial-grade printers and substances require high initial funding. In keeping with component price, mass production is more expensive than injection molding or casting.
6. Skills gap and standardization: Specialized expertise is required for the layout, operation, and upkeep of AM systems. Loss of generic standards and certification hints for a nice warranty.
7. Intellectual property concerns: Virtual files can be effortlessly copied, leading to IP theft and counterfeit production risks.

At the same time, AM has demonstrated its value in specialized and high-performance applications, and its business adoption is currently at its maximum, where complexity, customization, and low-volume manufacturing are key drivers. With the mixing of AI and Industry 4.0 technologies, many of these drawbacks may be addressed, making AM extra aggressive and dependable for broader use instances (Zhang, H. et al., 2020).

1.3 Core Concepts of Industry 4.0

Industry 4.0, frequently called the Fourth Industrial Revolution, represents a transformative shift in manufacturing through the convergence of virtual technologies, automation, and data-driven intelligence. At the heart of this revolution lies the seamless integration of the physical and digital worlds, made viable by numerous foundational principles, one of the most vital being CPS.

1.3.1 Cyber-Physical systems

CPS are smart mechanisms that tightly combine computational (cyber) and physical methods via embedded systems, sensors, actuators, and real-time communication networks. In CPS, the hardware and software program components are deeply intertwined, constantly interacting with and adapting to each other through feedback loops. In manufacturing, CPS permits the tracking, control, and automation of physical techniques through virtual twins, sensors, edge computing, and machine learning, consequently forming the foundation of smart factories (Zhang, Y. et al., 2020).

Key components of CPS in manufacturing include the following:

- Sensors and actuators: Gather real-time information (e.g., temperature, vibration, and build status) from physical structures. It enables dynamic manipulation of equipment and techniques in response to changing situations, as shown in Figure 1.7.
- Embedded systems: Microcontrollers and processors embedded in gadgets and machines that execute and manipulate common sense locally.
- Conversation infrastructure: High-speed networks (e.g., 5G, business Ethernet) that join physical devices with virtual systems. It facilitates machine-to-machine and machine-to-human interactions.
- Computation and analytics: Use AI, data mining, and simulation algorithms to analyze data and generate insights. Real-time decision-making via on-premises or cloud computing.
- Human–machine interfaces: Dashboards, augmented reality/virtual reality structures, or cellular apps that allow human operators to interact with and oversee CPS.

1.3.1.1 Functions of CPS in AM

In AM, CPS is vital in permitting clever, adaptive, and self-optimizing manufacturing structures. Key contributions include:

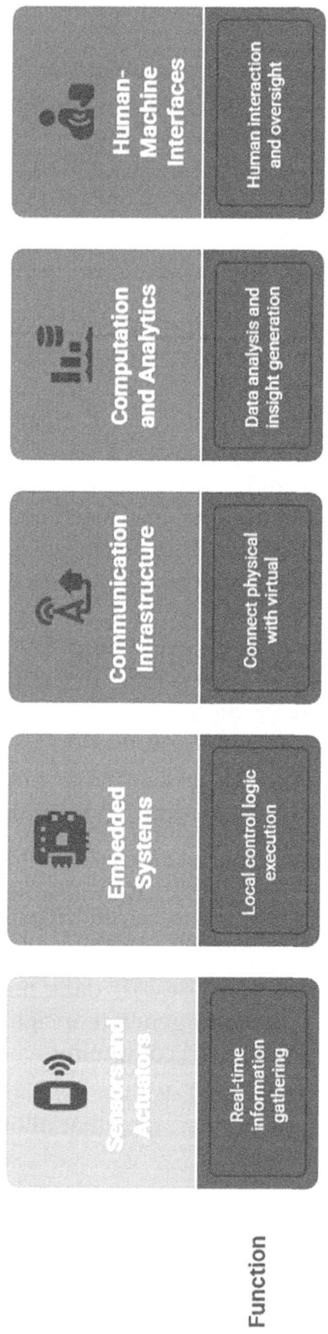

FIGURE 1.7
CPS in manufacturing components.

- Real-time tracking: Sensors embedded in 3D printers can tune nozzle temperature, cloth flow, layer alignment, and print bed conditions.
- Feedback loops: If anomalies, such as warping or delamination, are detected, CPS can routinely alter printing parameters like speed, temperature, or fabric feed fee.
- Predictive preservation: AI algorithms analyze usage information to identify when machine failures or maintenance requirements are expected before downtime happens.
- First-class control: CPS allows in-situ inspection and correction, enhancing consistency and reducing submit-processing.
- Records logging and traceability: Each print activity can be tracked from beginning to end, ensuring complete virtual documentation for certification and compliance.

1.3.1.2 Advantages of CPS in Industry 4.0-Enabled AM

- Extended autonomy: Machines could make localized selections based on information without human intervention.
- Improved performance: Optimized methods lessen waste, time, and electricity consumption.
- Customization at scale: CPS allows for the dynamic reconfiguration of printers to satisfy various client requirements in real time.
- Better protection: Computerized structures reduce human exposure to dangerous environments.

1.3.2 Internet of Things

IoT refers to the community of physical gadgets, including sensors, actuators, machines, and systems embedded with electronics, software, and connectivity that permit them to accumulate, trade, and act upon data over the net or business networks. Within the enterprise 4.0 context, IoT is a critical enabler of smart production, allowing machines and systems to be smart, self-aware, and interconnected. IoT transforms conventional production environments into digitally integrated ecosystems, wherein each factor constantly communicates and cooperates with others to improve performance, productivity, and agility (Zelinski, 2017).

1.3.2.1 Key Components of IoT in Manufacturing

- Smart Sensors: They measure and transmit real-time facts, including temperature, humidity, vibration, speed, pressure, and more.

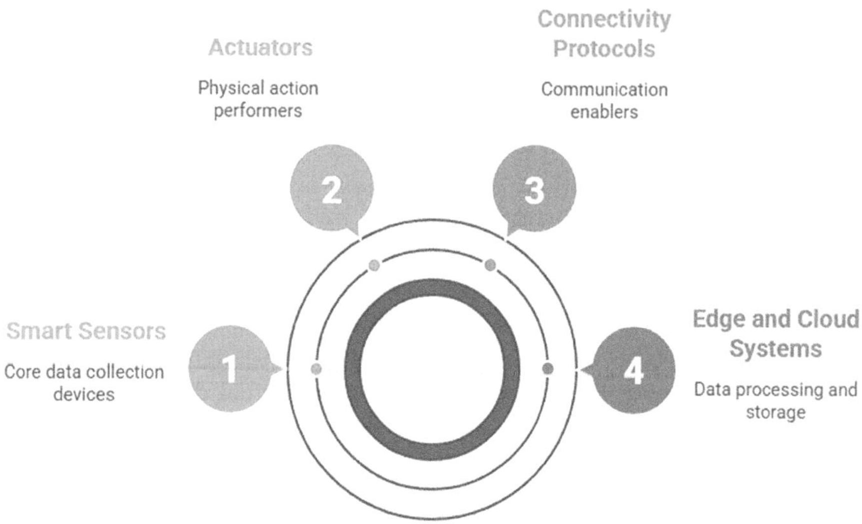

FIGURE 1.8
IoT components in manufacturing.

Examples: Thermocouples in print heads, load cells in construct platforms shown in Figure 1.8.

- Actuators: These perform physical actions based on instructions from control systems, such as adjusting nozzle height or pausing a print process.
- Connectivity protocols: These consist of Wi-Fi, Bluetooth, Zigbee, LoRa, and business Ethernet for seamless device communication.
- Edge and cloud systems: It permits local processing near the source, reducing latency. Cloud systems provide centralized statistics, analytics, and device orchestration.
- Records analytics and visualization: Dashboards, tracking structures, and AI algorithms provide actionable insights into device overall performance, product excellence, and predictive developments as shown in Figure 1.8.

1.3.2.2 Function of IoT in Additive Manufacturing

Combining IoT with AM complements visibility, traceability, and performance across the 3D printing lifecycle. Packages include:

- Real-time process tracking: IoT-enabled printers can relay real-time information about temperature gradients, build development, filament feed costs, and system fitness to centralized dashboards or mobile devices.

- Predictive maintenance: Sensors detect early symptoms of damage or bizarre vibrations, allowing predictive algorithms to schedule timely preservation and minimize downtime.
- Remote management and diagnostics: Operators can initiate, pause, or terminate print jobs remotely and diagnose mistakes without physical presence, permitting the dispensed and faraway manufacturing fashions shown in Figure 1.9.
- Inventory and material tracking: IoT systems track material usage and mechanically reorder supplies based on real-time consumption costs, ensuring continuous production.
- Quality assurance: Continuous monitoring ensures adherence to design specifications, with deviations flagged instantly for correction or reprinting (Zhou et al., 2015).

1.3.2.3 Advantages of IoT in Industry 4.0-Enabled AM

- Operational transparency: Stepped forward visibility into each layer of the AM method.
- Records-driven decisions: Actual-time analytics tell higher making plans, layout, and production techniques.
- Expanded performance: Optimized workflows, decreased material waste, and quicker bug detection.
- Remote accessibility: Complements flexibility and responsiveness in international production operations.
- Integration with AI: IoT statistics serves as gas for machine learning algorithms to power nonstop improvement.

1.3.2.4 Challenges and Issues

- Statistics security: Expanded connectivity introduces vulnerabilities that must be mitigated through robust cybersecurity practices.
- Interoperability: Ensuring seamless communication throughout numerous devices and platforms.
- Records overload: Efficient records management and filtering are crucial to avoid being beaten by noncrucial information.
- Initial investment: In-advance expenses for IoT infrastructure and skilled employees can be enormous.

IoT is a cornerstone of Industry 4.0, and its integration into AM procedures unlocks extraordinary stages of automation, perception, and adaptability. With the aid of connecting devices, structures, and people into a unified digital atmosphere, IoT lays the foundation for smart factories, which can

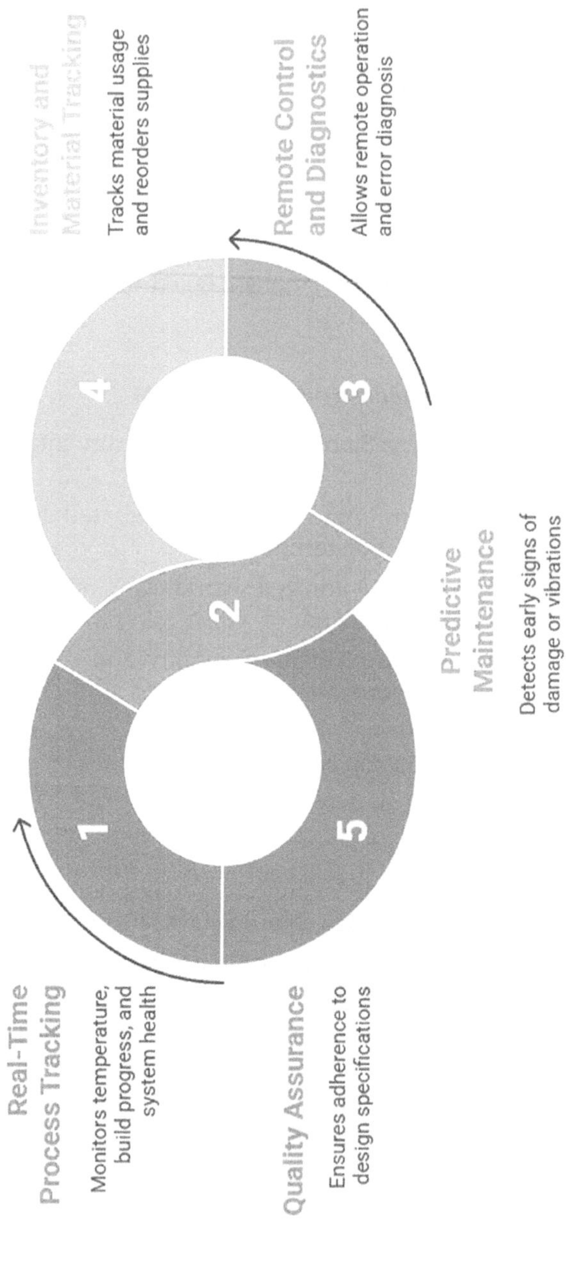

FIGURE 1.9
IoT in additive manufacturing cycle.

be self-aware, agile, and capable of self-optimization. As AM evolves, IoT will continue to be a crucial device for ensuring resilience and innovation in manufacturing systems.

1.3.3 Big Data Analytics, Cloud Computing, and Edge AI

The fusion of big data analytics, cloud computing, and edge AI forms a critical triad within the enterprise 4.0 ecosystem, especially in enhancing AM's intelligence, velocity, and performance. As AM tactics generate vast volumes of heterogeneous information from temperature logs and material usage to 3D version revisions and device behavior, these technologies enable smart facts usage for real-time decision-making and long-term optimization (Zhu et al., 2022).

1.3.3.1 Big Data Analytics in AM

This analytics refers to the use of advanced techniques, consisting of machine learning, statistical modeling, and predictive analytics – to derive actionable insights from large, complicated datasets. In AM, this consists of structured and unstructured records from CAD documents, sensor feeds, renovation logs, and manufacturing effects.

1.3.3.1.1 Use of Big Data in AM

- Method optimization: Analyzing print parameters (e.g., temperature, velocity, and layer top) to optimize builds for power, precision, and cloth performance.
- Disorder prediction: Identifying patterns and correlations that cause defects, together with warping, voids, or layer shifting.
- Lifecycle analytics: Tracing components from layout to put up-processing to understand long-term overall performance and disasters.
- Client personalization: Leveraging person statistics to customize designs and functionality at scale.

1.3.3.1.2 Benefits

a. Stepped forward product best and consistency
b. Decreased trial-and-error in component improvement
c. Predictive modeling for yield and throughput

1.3.3.2 Cloud Computing in AM

Cloud computing affords on-demand access to computational control, storage, and software applications over the network. In AM, the cloud is a crucial hub for collaboration, data storage, AI version deployment, and workflow orchestration.

1.3.3.2.1 Programs in AM

- Layout sharing and collaboration: Groups can access, edit, and validate 3D models globally via cloud-based CAD platforms.
- Remote tracking and control: Operators can monitor and manipulate AM machines and production strategies from anywhere.
- AI-as-a-service: Cloud platforms offer scalable AI gear for best management, simulation, and layout automation.
- Information archival and compliance: Cloud answers assist traceable information storage for regulatory and certification functions.

1.3.3.2.2 Benefits

- Scalability and decreased infrastructure price.
- Real-time international collaboration.
- Seamless integration with IoT and AI packages.
- Faster deployment of the system, getting to know the pipelines.

1.3.3.3 Edge AI in AM

Edge AI combines AI and edge computing, permitting statistics to be processed regionally on gadgets (such as printers or sensors) rather than sent to the cloud. This enables low-latency, self-sustaining decision-making, an essential requirement for time-sensitive production operations.

1.3.3.3.1 Applications in AM

- Anomaly detection: Edge devices can halt or adjust printing immediately upon detecting real-time deviations (e.g., nozzle clogging or overheating).
- Adaptive printing: AI fashions deployed at the edge can dynamically regulate parameters mid-print primarily based on live data remarks.
- Offline intelligence: Isolated or disconnected environments can nevertheless gain from AI without counting on cloud access.
- Stronger security: Sensitive records remain nearby, decreasing cybersecurity threats.

1.3.3.3.2 Benefits

- Real-time responsiveness and self-sustaining manipulation.
- Reduced network dependency and bandwidth usage.
- Superior reliability in decentralized manufacturing environments.
- Quicker comment loops for nonstop, first-rate assurance.

Combining big data, cloud computing, and edge AI into AM exemplifies the essence of Industry 4.0-related, shrewd, and adaptive manufacturing surroundings. These technologies empower producers with deep operational insights and force faster innovation, improved product quality, and increased customer responsiveness. As AM becomes extra-complicated and records-intensive, this virtual trifecta could be instrumental in unlocking its complete business capacity.

1.4 Role of Artificial Intelligence in AM

AI is revolutionizing the way products are conceived, designed, and manufactured. Within the context of AM, AI offers advanced talents for automating decision-making, improving performance, and enabling designs that have been previously unattainable with the use of traditional strategies. Among its many applications, design optimization powered through generative layout and topology optimization stands as one of the most impactful ones (Shafiee, 2025).

1.4.1 AI in Design Optimization (e.g., Generative Design and Topology Optimization)

Design optimization refers to the process of improving an element's or structure's geometry to meet precise performance standards while minimizing application use. When blended with AM, AI-enabled optimization unlocks new possibilities in creating lightweight, sturdy, and material-efficient additives that are regularly too complicated to manufacture using subtractive strategies.

Distinguished AI-pushed strategies in this area are:

a. Generative Design.
b. Topology Optimization.

1.4.1.1 Generative Design

It is an AI-powered method in which the software program explores an enormous solution space to automatically generate a couple of design alternatives based totally on predefined constraints and targets (e.g., weight, energy, cost, material, and manufacturing technique) (Singh et al., 2021).

Workflow

- The engineer inputs performance requirements (e.g., load situations, material kind, and allowable space).

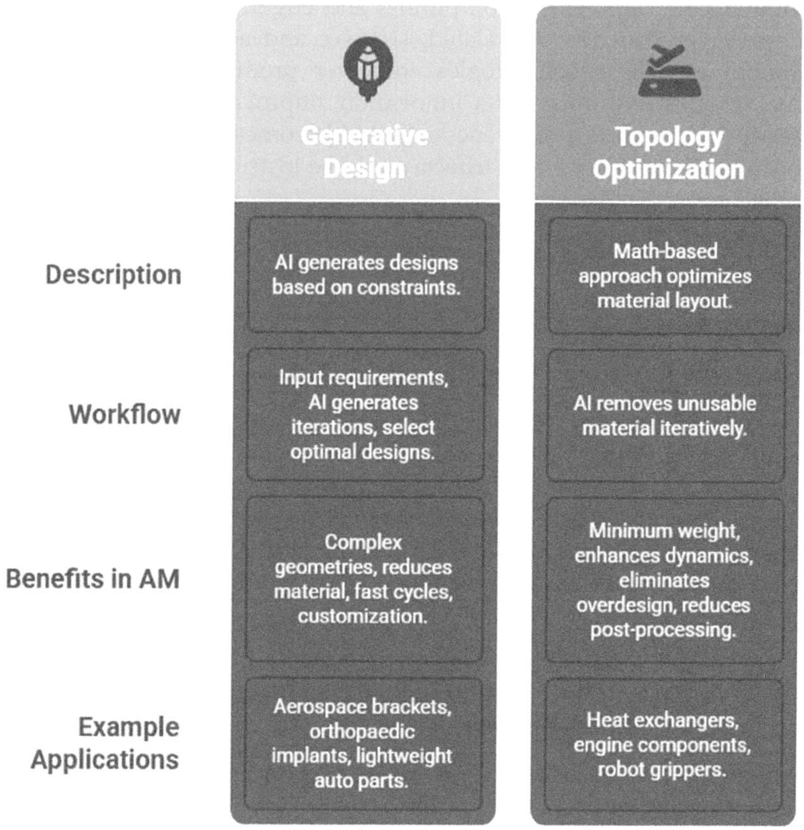

FIGURE 1.10
AI-driven design optimization techniques.

- AI algorithms generate lots of valid layout iterations.
- The most beneficial designs are selected for simulation and production, as shown in Figure 1.10.

1.4.1.1.1 Advantages in AM

- Produces exceedingly organic and complex geometries best suited for 3D printing.
- Reduces material utilization and element weight without compromising control.
- Fastening innovation cycles through computerized design exploration.
- Permits customization at scale for biomedical, aerospace, and automobile applications.

1.4.1.1.2 *Instance Applications*

- Aerospace brackets with inner lattice systems.
- Custom orthopedic implants designed for bone integration.
- Lightweight automobile additives with optimized strain distribution.

1.4.1.2 *Topology Optimization*

This is a mathematical approach that determines the high-quality material design within a given layout space for a selected set of loads, boundary conditions, and constraints.

AI and simulation algorithms iteratively dispose of unusable materials from a shape while keeping performance objectives. This creates voids, inner channels, or nonlinear contours, which can be perfect for fabrication via additive manufacturing (Thompson et al., 2016).

1.4.1.2.1 *Benefits in AM*

- Achieves minimum weight with most stiffness or energy.
- Enhances thermal or fluid dynamics in functionally optimized components.
- Eliminates overdesign with the aid of focusing material most effectively wherein needed.
- Reduces submit-processing and assembly necessities by using allowing part consolidation.

1.4.1.2.2 *Instance Applications*

- Warmth exchangers with optimized thermal float paths.
- Engine components with minimal mass for fuel performance.
- Robot grippers with power-centered structural integrity.

1.4.1.3 *Synergy with Machine Learning*

AI algorithms along with neural networks, genetic algorithms, and reinforcement learning are being incorporated into optimization tools to enhance prediction accuracy and answer convergence:

- Neural networks are expecting overall performance metrics based on geometric functions.
- Genetic algorithms evolve design options via choice and mutation strategies.
- Reinforcement learning helps optimize designs over sequential decisions.

1.4.1.4 Challenges and Considerations

- Computational complexity: Optimization algorithms are computationally intensive and require effective hardware or cloud support.
- Layout interpretation: The resulting organic structure may be difficult to interpret or validate when using conventional CAD tools.
- Manufacturability: A few optimized designs can also nevertheless require submit-processing or support structure concerns (Ucar et al., 2024).

AI-driven design optimization is revolutionizing the way engineers approach product development in additive manufacturing. By using generative design and topology optimization, manufacturers can create high-performance, lightweight, and cost-effective components that take full advantage of the capabilities of additive manufacturing. Those AI strategies not only simply reduce time and material waste but also lead to groundbreaking designs that redefine the boundaries of engineering.

1.4.2 AI in Process Parameter Tuning

It is an essential element of accomplishing objectives in AM. Parameters that include laser power, scanning speed, layer thickness, nozzle temperature, and build orientation without delay impact the component's structural integrity, dimensional accuracy, and surface finish. However, manually optimizing these parameters is difficult, time-consuming, and frequently inconsistent. This is where AI performs a transformative role, automating and enhancing the tuning procedure through proven methods methods (Khajavi et al., 2014).

1.4.2.1 Role of AI in Parameter Optimization

AI allows the smart choice, prediction, and optimization of system parameters based on historical facts, real-time data, and simulation effects. By gaining knowledge of models, especially supervised studying, reinforcement learning, and optimization algorithms, it can find the most advantageous parameter units that maximize component quality while minimizing defects and manufacturing time.

1.4.2.2 Key AI Techniques for Parameter Tuning

- Supervised ML: Trained on categorized datasets (e.g., parameters and corresponding effects). It predicts superior parameter combos for given part geometries or materials.

Standard algorithms: Decision tree, random forests, support vector machine, neural networks.

- Reinforcement Learning: Learns through trial and error by interacting with the environment. Ideal for real-time adaptive manipulation of system parameters.

Variables can be regulated with laser power or feed charge throughout the construction procedure to compensate for anomalies.

- Bayesian optimization efficiently explores the design space to identify the good parameter sets with minimal experimentation. This technique is useful in high-dimensional and costly-to-compare issues, such as those in metallic AM.

1.4.2.2.1 *Applications in AM*

- Steel Powder Bed Fusion (e.g., SLM and DMLS): Optimizing laser experiment strategies, layer thickness, and power for minimizing porosity and cracking.
- FDM: Adjusting print pace, nozzle temperature, and infill patterns for better floor finish and electricity as per Figure 1.11.
- SLA: Controlling exposure time and curing parameters ensures dimensional constancy and solid properties (Li et al., 2021).

1.4.2.2.2 *Benefits of AI-Based Parameter Tuning*

- Improved element eminence: Stronger dimensional accuracy, mechanical power, and surface finish.
- Reduced errors: Minimizes material waste, electricity usage, and experimental cycles.
- Quicker time-to-market: Streamlined manufacturing cycles with quicker setup and less screwups.
- Scalability: Allows steady tuning across multiple machines or production environments.
- Customization: Automatically adapts parameters for varying geometries, substances, and environmental situations (Hernandez Korner et al., 2020).

1.4.2.2.3 *Case Examples*

- Aerospace: AI models skilled on ancient builds can right away expect the proper laser parameters for brand spanking new geometries, decreasing the number of qualification checks.

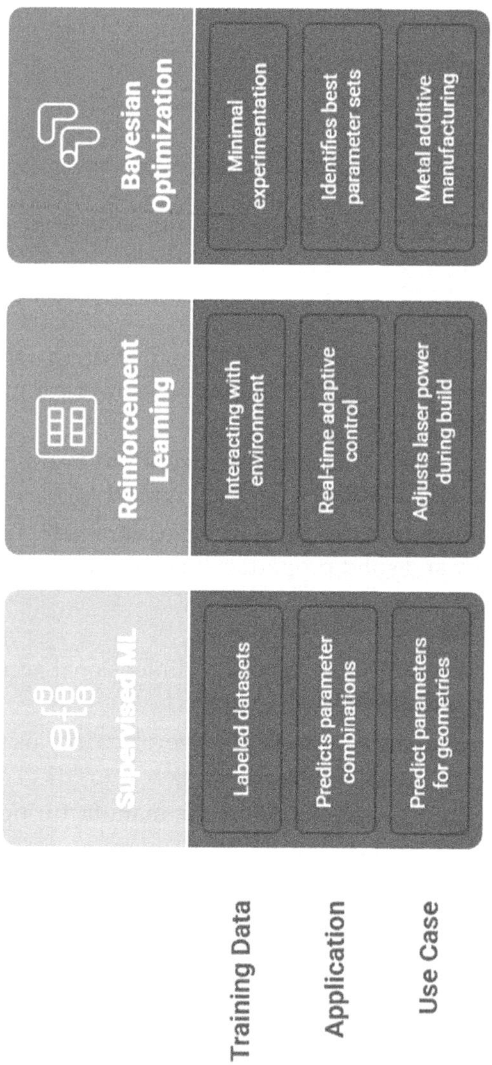

FIGURE 1.11
AI techniques in additive manufacturing.

- Biomedical: AI tunes print settings for biocompatible substances in implants, optimizing each structural and organic properties.
- Automotive: Reinforcement studying dynamically controls print conditions to keep high quality on long-length builds.

1.4.2.2.4 Challenges and Considerations

- Data dependency: Requires labeled datasets for effective version training.
- Version generalization: Ensuring AI fashions are sturdy across extraordinary machines and situations.
- Integration with hardware: Real-time AI facilitates the coordination and integration of sensors and machine controllers.

AI-pushed method parameter tuning is a cornerstone of shrewd additive manufacturing. By enabling precise, adaptive, and automated manipulation over vital method variables, AI improves product performance and consistency and significantly reduces development effort and production expenses. As AM systems continue to evolve, mixing AI for parameter optimization might be crucial for reaching scalable and customized manufacturing.

1.4.3 AI in Quality Control and Maintenance in AM

Quality control and preservation are essential for AM techniques' reliability, safety, and efficiency. Conventional high-quality warranty techniques, which depend heavily on guided inspections and submit-production testing, are time-consuming and prone to human errors. Integrating AI, especially ML and deep learning (DL), enables real-time, excellent tracking, illness detection, and predictive preservation. These technologies permit for smarter, self-correcting structures that beautify productivity and decrease downtime (Gao et al., 2015).

1.4.3.1 Predictive Maintenance and Failure Detection

This uses AI to assume gadget failures before they occur by reading historical and real-time records from gadget sensors (e.g., vibration, temperature, and acoustic emissions). This permits proactive interventions that prevent sudden breakdowns and increase equipment lifespan.

Key features

- Sensor fusion: AI aggregates statistics from multiple resources (e.g., vehicles, extruders, and laser modules) for comprehensive evaluation.
- Anomaly detection: ML fashions hit upon deviations from ordinary running styles that may imply mechanical wear or failure onset.

- Failure prediction models: Time-collection forecasting and class algorithms expect probably failure factors and their root causes.

Advantages

- Minimizes unscheduled downtime and manufacturing interruptions.
- Reduces preservation charges via optimized servicing schedules.
- Will increase average system usage and reliability.

Machine Learning and Deep Learning for AM Quality Control

These kinds of learning have emerged as recreation-changers for great assurance in AM, imparting automated, accurate, and nondestructive evaluation techniques (Lu, 2017).

- ML techniques

  Supervised knowledge is trained on labeled datasets of excellent and faulty parts to classify future prints.

  Unsupervised knowledge of clusters records to perceive outliers or novel illness patterns without needing labels.

  The system learns the most effective manipulation strategies through continuous feedback and reward mechanisms (Hendrixson, 2017).
- DL Techniques

  Convolutional neural networks examine photograph records (from cameras or thermal sensors) to identify surface defects, porosity, layer delamination, and warping.

  Recurrent neural networks: Used for sequence modeling of time-collection sensor facts to detect method instabilities as shown in Figure 1.12.

  Autoencoders learn compressed representations of regular procedure conduct and flag deviations as capacity defects.

Applications

- Visual inspection of printed layers.
- Class of microstructural anomalies in metallic AM.
- In-situ defect prediction using thermal or acoustic information.

AI is adapting toward eminence and maintenance in AM, from reactive and manual techniques to proactive, automatic, and intelligent systems. Via predictive protection, system-studying-powered defect detection, and real-time remarks loops, AI now complements the reliability of AM equipment and the consistency and quality of the parts produced. As these technologies mature, they may be pivotal in making AM more scalable, price-competitive, and production-ready for vital commercial applications (Mourtzis et al., 2018).

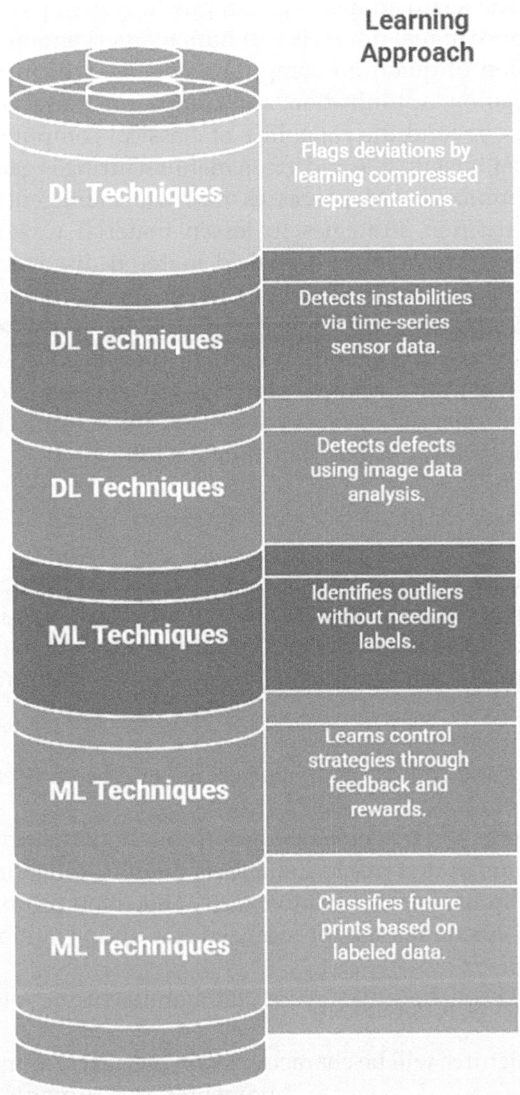

FIGURE 1.12
Machine learning techniques spectrum for AM quality control.

1.5 Future Trends and Research Directions

The future of AM lies in the convergence of advanced AI technology and next-era computing paradigms to create independent, intelligent, and sustainable systems. One rising trend is the development of self-sufficient and self-learning AM structures that leverage reinforcement learning and closed-loop feedback to adaptively manipulate system parameters and optimize performance in real time without human intervention. Another frontier is the integration of quantum computing and AI, which promises to revolutionize simulation, optimization, and complex material conduct modeling by fixing issues beyond the capability of classical computers (Budzik et al., 2022). In parallel, sustainable and green manufacturing is gaining momentum through AI-optimized AM processes that use intelligently designed and executed construction strategies to lessen material waste, energy intake, and emissions. Additionally, AI-pushed material discovery is accelerating the identification and improvement of novel alloys, composites, and smart materials tailor-made for AM applications, utilizing machine learning to predict fabricability and printability with high precision. These developments together signal a transformative shift toward smarter, cleaner, and more efficient additive manufacturing ecosystems. Some future trends in AI and Industry 4.0 integration with AM include the following.

- Digital twins for predictive simulation and control

The use of virtual twins' digital replicas of physical AM structures is expected to become more advanced. These twins will constantly update with real-time sensor statistics, enabling advanced simulations, predictive best manipulation, and automatic decision-making to pre-emptively address deviations in the printing process.

- Federated and collaborative mastering in AM ecosystems

As more than one AM machine operates throughout distributed factories or deliver chains, federated mastering allows them to collaboratively improve AI fashions without sharing sensitive facts. This trend complements privacy, scalability, and customization for localized AM optimization.

- Human-AI collaborative layout and manufacturing interfaces

Future AM structures will be characterized by intuitive, immersive interfaces wherein people and AI co-create. Augmented reality, natural language processing, and explainable AI will assist engineers in interacting with complex AI-driven design and production recommendations more transparently and efficiently.

- AM-as-a-service structures with smart scheduling

Cloud-based AM offerings will evolve into smart structures that offer layout-to-production pipelines. These structures will leverage AI for wise scheduling, value estimation, material choice, and automatic compliance exams, permitting corporations to successfully order custom-designed elements without owning infrastructure.

- Integration of blockchain for relaxed and transparent AM tactics

As AM will become mainstream in regulated sectors like aerospace and healthcare, blockchain can be integrated to ensure traceability, IP protection, and comfortable sharing of design documents and method parameters. AI will help track and validate the integrity of transactions and statistics across decentralized AM networks.

1.6 Conclusions

The convergence of AI, Industry 4.0 technology, and AM is revolutionizing the manufacturing landscape by allowing intelligent, adaptive, and green manufacturing systems. This chapter has highlighted key improvements consisting of AI-pushed design optimization, process parameter tuning, predictive upkeep, and real-time excellent control, all of which enhance precision, reduce waste, and enhance scalability in AM methods. The combination of technologies like IoT, CPS, cloud computing, and big data analytics further strengthens the digital thread throughout the manufacturing lifecycle. As the sector evolves, embracing autonomous systems, sustainable practices, and AI-guided material discovery might be necessary. Researchers are encouraged to discover interdisciplinary frameworks that combine AI, quantum computing, and digital twins, even as practitioners should prioritize scalable AI answers, relaxed data infrastructures, and workforce reskilling. Collectively, these efforts will impact the following generation of clever, sustainable, and resilient additive production ecosystems.

References

Bendoly, E., Chandrasekaran, A., Handfield, R., Khajavi, S. H., Roscoe, S., & Ferreira Lima, M. D. R. (2024). The role of generative design and additive manufacturing capabilities in developing human–AI symbiosis: Evidence from multiple case studies. *Decision Sciences*, 55(4), 325–345. https://doi.org/10.1111/deci.12619

Budzik, G., Dziubek, T., & Słota, J. (2022). AI-based support system for monitoring the quality of a product manufactured by additive manufacturing. *Applied Sciences*, 12(22), 11345. https://doi.org/10.3390/app122211345

Chen, Y., et al. (2022). AI-based quality control in additive manufacturing: A review. *Journal of Intelligent Manufacturing*, 33(2), 1–15. https://doi.org/10.1007/s10845-021-01765-4

Dehghan, S., Sattarpanah Karganroudi, S., Echchakoui, S., & Barka, N. (2025). The integration of additive manufacturing into Industry 4.0 and Industry 5.0: A bibliometric analysis (trends, opportunities, and challenges). *Machines*, 13(1), 62. https://doi.org/10.3390/machines13010062

Gao, W., et al. (2015). The status, challenges, and future of additive manufacturing in engineering. *Computer-Aided Design*, 69, 65–89. https://doi.org/10.1016/j.cad.2015.04.001

Gupta, A., et al. (2021). AI-enabled predictive maintenance in manufacturing: A review. *Journal of Manufacturing Systems*, 59, 1–13. https://doi.org/10.1016/j.jmsy.2021.01.002

Hendrixson, S. (2017). How to think about design for additive manufacturing. *Modern Machine Shop*. www.mmsonline.com/articles/how-to-think-about-design-for-additive-manufacturing

Hernandez Korner, M. E., Lambán, M. P., Albajez, J. A., Santolaria, J., Ng Corrales, L. d. C., & Royo, J. (2020). Systematic literature review: Integration of additive manufacturing and Industry 4.0. *Metals*, 10(8), 1061. https://doi.org/10.3390/met10081061

Khajavi, S. H., Partanen, J., & Holmström, J. (2014). Additive manufacturing in the spare parts supply chain. *Computers in Industry*, 65(1), 50–63. https://doi.org/10.1016/j.compind.2013.07.008

Khorasani, M., Loy, J., Ghasemi, A. H., Sharabian, E., Leary, M., Mirafzal, H., Cochrane, P., Rolfe, B., & Gibson, I. (2022). A review of Industry 4.0 and additive manufacturing synergy. *Rapid Prototyping Journal*, 28(8), 1462–1475. https://doi.org/10.1108/RPJ-08-2021-0194

Kumar, A., et al. (2019). Additive manufacturing applications in Industry 4.0: A review. *Journal of Manufacturing Systems*, 50, 1–14. https://doi.org/10.1016/j.jmsy.2019.01.006

Lee, J., et al. (2014). Predictive maintenance of industrial systems using big data analytics. *Journal of Manufacturing Science and Engineering*, 136(6), 061005. https://doi.org/10.1115/1.4026360

Li, X., et al. (2021). Deep learning in additive manufacturing: A review. *Journal of Manufacturing Processes*, 68, 1–13. https://doi.org/10.1016/j.jmapro.2021.05.003

Lu, Y. (2017). Industry 4.0: A survey on technologies, applications and open research issues. *Journal of Industrial Information Integration*, 6, 1–10. https://doi.org/10.1016/j.jii.2016.04.001

Mourtzis, D., et al. (2018). The potential of additive manufacturing in the smart factory: A review. *Applied Sciences*, 9(18), 3865. https://doi.org/10.3390/app9183865

Sani, M. A., et al. (2024). Artificial intelligence-augmented additive manufacturing: Insights on closed-loop 3D printing. *Advanced Intelligent Systems*, 6(5), 2400102. https://doi.org/10.1002/aisy.202400102

Shafiee, S. (2025). Generative AI in manufacturing: A literature review of recent applications and future prospects. *Procedia CIRP*, 132, 1–6. https://doi.org/10.1016/j.procir.2025.01.001

Singh, R., et al. (2021). Design for additive manufacturing: A systematic review. *Sustainability*, 13(19), 7936. https://doi.org/10.3390/su13197936

Tang, Y., et al. (2016). A survey of the design methods for additive manufacturing to improve functional performance. *Rapid Prototyping Journal*, 22(3), 569–590. https://doi.org/10.1108/RPJ-12-2014-0171

Thompson, M. K., et al. (2016). Design for additive manufacturing: Trends, opportunities, considerations, and constraints. *CIRP Annals*, 65(2), 737–760. https://doi.org/10.1016/j.cirp.2016.05.004

Ucar, A., Karakose, M., & Kırımça, N. (2024). Artificial intelligence for predictive maintenance applications: Key components, trustworthiness, and future trends. *Applied Sciences*, 14(2), 898. https://doi.org/10.3390/app14020898

Varela, M. L. R., Araújo, A., & Ferreira, L. P. (2018). Additive manufacturing at Industry 4.0: A review. *International Journal of Engineering and Technical Research*, 8(1), 1–5.

Wang, L., et al. (2023). Real-time monitoring in additive manufacturing: A review. *Additive Manufacturing*, 46, 102109. https://doi.org/10.1016/j.addma.2021.102109

Zelinski, P. (2017). GE team secretly printed a helicopter engine, replacing 900 parts with 16. *Modern Machine Shop*. www.mmsonline.com/articles/ge-team-secretly-printed-a-helicopter-engine-replacing-900-parts-with-16

Zhang, H., et al. (2020). Machine learning in additive manufacturing: A review. *Journal of Manufacturing Science and Engineering*, 142(11), 110801. https://doi.org/10.1115/1.4047357

Zhang, Y., et al. (2020). Artificial intelligence for additive manufacturing: A review. *Advanced Intelligent Systems*, 2(7), 2000030. https://doi.org/10.1002/aisy.202000030

Zhou, K., Liu, T., & Zhou, L. (2015). Industry 4.0: Towards future industrial opportunities and challenges. In *2015 12th International Conference on Fuzzy Systems and Knowledge Discovery (FSKD)* (pp. 2147–2152). IEEE. https://doi.org/10.1109/FSKD.2015.7382284

Zhu, Y., et al. (2022). Predictive maintenance and intelligent sensors in smart factories. *Sensors*, 22(3), 1234. https://doi.org/10.3390/s22031234

2

Predicting Productivity and Surface Quality in EN19 Alloy Electrical Discharge Machining Using Machine Learning Regression Models

Neeta Deshpande, Vijaykumar S. Jatti, A. Saiyathibrahim, Abhinav Kumar, Shyan, Pallavi Sharma, Patel Amishkumar Bhagvandas, and Sergey E. Shcheklein

2.1 Introduction

The impact of artificial intelligence (AI) on manufacturing has made traditional machining techniques, such as those in electrical discharge machining (EDM), more sustainable. Machine learning forecasts productivity outcomes and the EDM surface quality of EN19 alloy, enhancing the sustainability of EDM and aiding in process optimization. Now in practice, eco-friendly manufacturing, which means the creation of reliable prediction models in order to choose a process with an environmentally friendly process setting, becomes even more crucial. This chapter utilizes advanced machine learning algorithms to exactly predict both productivity and surface quality with reduced energy, less resource waste, and a more efficient process. Using advanced regression analysis, the study finds relationships between EDM factors such as pulse-on time, gap current, and tool diameter and the positive environmental effects of the process. By doing this, we increase the development of smart manufacturing and facilitate the industrial processes to become more sustainable (Kaur et al., 2025; Ishfaq et al., 2024). Breaking down difficult materials in our daily processes is a common tool of the industrial sector, one being EDM (Kamenskikh et al., 2023). The researchers discover that EDM surface quality and productivity, in general, present many difficulties in further improvement. Investigations in different fields were carried out to find solutions based on how a number of input factors control the material removal rate (MRR) and surface roughness (Ra) (Jain & Parashar 2021).

 DOI: 10.1201/9781003506492-2

Scientists have analyzed the adjustments in parameters to better understand their influences on MRR alongside surface roughness development. The central composite experimental design helped create a polynomial MRR model during AISI D2 tool steel machining operations with copper electrodes (Pradhan & Biswas 2011). Research on EN-19 alloy steel MRR using EDM employs response surface methodology (RSM) for analyzing multiple process parameter interactions (Surekha et al., 2021). Researchers optimized the multi-responses of SLMed AlSi10Mg in Wire-EDM using Taguchi integrated Grey Relational Analysis (GRA) to achieve maximum MRR and minimum SR. Testing confirmed optimal parameter settings delivered a 38.57% boosted system performance (Ranganathan et al., 2025). Different combinations of positive polarity and extended pulse length resulted in thicker recast layer formation in Ni-based super alloys based on Taguchi research method (Upadhyay et al., 2017). The experimental design method from Taguchi illuminated how MRR enhancement together with improved surface roughness characterizations are positively linked to Ton and current control factors (Sharma et al., 2021). Using the Adaptive Neuro-Fuzzy Inference System to evaluate tool wear rate (TWR), it was shown that input current and pulse on time were the most influential key factors (Bhiradi et al., 2020). The implementation of RSM to study the effects of Ton, V, and Toff on EDM performance measures using analysis of variance was done to recognize individual process variables (Sharma et al., 2022). L18 Taguchi experimental design used a study with differernt electrode shapes (triangle, square, and circular) and found that it affected NiTi SMA's EDM process parameters when using graphite tools (Gaikwad & Jatti 2018). EDM process has been widely utilized in various machining procedures. SR, however, is a crucial component that influences the tensile strength, wear resistance, fatigue strength, and ductility of machine components in real-world applications (Qudeiri et al., 2020). Machining Al 6061 alloy using nanopowder-mixed EDM, which tackles the issues of high electrode wear rate (EWR) and overcutting (OC), was studied through artificial neural networks (ANN) and multi-objective genetic algorithms. The use of cryogenic-treated brass electrodes in deionized water notably enhances the reduction of EWR and OC (Sana et al., 2024). Optimization of EDM parameters for Cu-based shape memory alloys using machine learning techniques such as genetic algorithm and teacher learning-based optimization (TLBO)) focused on the effects of pulse-off time, pulse-on time, peak current, and gap voltage on dimensional deviation and TWR (Singh et al., 2022).

Based on the survey or previous works, the uniqueness of this study lies in its application of machine learning regression models to predict productivity and surface quality in EDM of EN19 alloy. While traditional methods typically rely on experimental data and empirical formulas, this study introduces advanced predictive models that utilize machine learning to enhance the accuracy and reliability of performance predictions. The research gap addressed is the limited use of machine learning techniques in EDM optimization for EN19 alloy, which is crucial for industries such as automotive and aerospace, where high precision is required. By leveraging machine learning, this study aims to fill the gap in real-time, data-driven predictions for improving machining efficiency and surface finish quality, offering a more sophisticated and efficient approach compared to conventional methods. This study investigated the behavior of different parameters on surface roughness, TWR, and material removal rate during the EDM of EN19 workpiece. Furthermore, six different ML regression algorithms such as multiple linear regression, Poisson's regression, polynomial linear regression, support vector regression, decision tree regression, and random forest regression are used to predict the best responses.

2.2 Materials and Methods

This study employed a die-sink EDM machine, specifically the model C 400×250 manufactured by Electronica Machine Tools Limited. It featured negative polarity electrodes and Z-axis NC control, as shown in Figure 2.1. The dielectric fluid used was a commercial-grade oil, with a side flushing pressure of 0.5 kg/cm^2. The workpiece material was EN 19 alloy steel, and the tool electrode was made of electrolytic copper. The study utilized tool electrodes with five different diameters (12, 15, 18, 21, and 24 mm), each measuring 50 mm in length. The EDM input parameters included current (4, 6, 8, 10, and 12 A), gap voltage (75, 65, 55, 45, and 35 V), pulse-on time (96, 73, 50, 27, and 4 μs), and pulse-off time (8, 7, 6, 5 and 4 μs). The response table for TWR, MRR, and Ra is presented in Table 2.1. The dataset, after removing the run number column, included values for current, voltage, electrode diameter, pulse-off time, and pulse-on time. These parameters were used to calculate MRR, TWR, and Ra using regression algorithms. The data was divided into training and testing datasets, with the training values used to practice the regression and the testing data utilized to evaluate its accuracy. Figure 2.2 illustrates the implementation of these algorithms, which involved obtaining a correlation heat map to understand the relationship between the EDM process parameters and the response variables. The dataset was partitioned into 70% training and 30% testing data, and hyperparameters were optimized to ensure the model's adequacy. The machine learning model performance assessment leveraged mean squared error (MSE) together with root mean squared error (RMSE), mean absolute error (MAE), and determination coefficient as metrics (Jatti et al., 2024).

FIGURE 2.1
Experimental setup used.

2.3 Results and Discussion

The EDM process of EN19 alloy displayed changing patterns for both TWR and MRR and surface roughness. The authors found that increased gap currents generated better material removal rates, yet high-density electron impingement produced variations in the TWR. The open-circuit voltage was observed to have a direct impact on MRR and Ra, with an enhancement in voltage resulting in higher MRR and surface roughness. Pulse-on time was proportional to the amount of energy used, with longer pulse durations improving MRR but leading to a rougher surface. Larger tool diameters resulted in greater surface area and higher MRR, but also increased surface roughness. The effects of various EDM parameters on MRR, TWR, and Ra were visualized using a correlation heat-map (Figures 2.3–2.5), which revealed that gap current had the highest correlation with all three parameters.

TABLE 2.1

Experimental Trials with Conditions and Observed Values

Run No.	Current I (A)	Voltage V	Ton (µs)	Toff (µs)	Tool	Run No.	Current I (A)	Voltage V
8	8	55	50	6	18	17.656	0.223	8.150
32	10	65	27	7	15	34.140	1.672	10.410
9	6	45	27	7	15	14.688	0.401	6.280
26	10	45	27	5	15	29.095	0.803	8.090
25	8	55	50	8	18	18.522	0.089	6.840
11	8	55	50	6	24	18.624	0.089	10.010
15	8	55	50	6	18	17.733	0.111	8.360
29	10	65	27	5	21	25.456	1.468	7.960
22	6	65	73	5	21	9.087	0.037	9.020
16	6	65	27	7	21	12.059	0.669	7.170
1	6	45	73	7	21	5.596	0.178	6.430
14	6	65	27	5	15	13.350	0.123	6.090
18	8	55	50	6	18	19.388	0.089	9.560
24	8	55	50	4	18	18.471	0.468	10.020
12	6	45	73	5	15	1.771	0.045	1.760
13	8	35	50	6	18	14.369	0.178	6.980
19	10	45	73	7	15	19.350	0.134	7.550
27	8	55	50	6	18	20.127	0.156	10.580
7	10	45	73	5	21	18.195	0.102	7.480
5	8	55	96	6	18	13.146	0.223	4.880
28	8	55	50	6	18	17.325	0.167	7.960
10	4	55	50	6	18	3.006	0.156	3.230
6	8	75	50	6	18	22.420	0.089	10.130
20	8	55	50	6	12	15.083	0.089	6.620
4	6	45	27	5	21	14.140	0.087	6.540
2	10	65	73	5	15	26.102	0.279	10.850
17	8	55	4	6	18	14.975	0.203	5.560
30	8	55	50	6	18	20.255	0.123	9.210
3	10	65	73	7	21	28.280	0.100	10.810
23	6	65	73	7	15	8.611	0.167	5.400
31	12	55	50	6	18	35.592	0.357	10.880
21	10	45	27	7	21	32.378	1.115	9.360

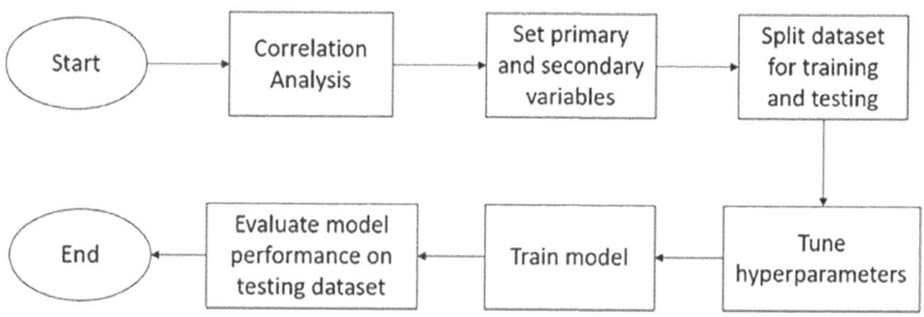

FIGURE 2.2
Machine learning regression model implementation workflow.

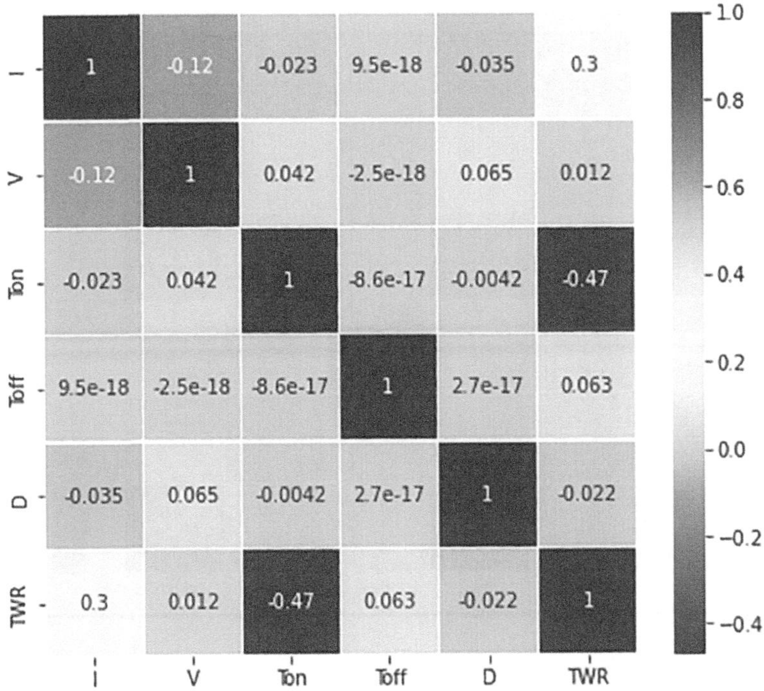

FIGURE 2.3
Correlation between input values and TWR.

The dataset reveals that pulse-off time has a negative correlation with MRR, TWR, and Ra. Notably, both current and voltage exhibit a significant impact on the values of MRR, TWR, and Ra. Based on this information, several regression models were applied to the dataset. Figures 2.6–2.11 illustrate the performance of various regression models with reference to metrics. The following regression algorithms were applied as part of the experiment to identify the outcome of the response variables.

2.3.1 Multiple Linear Regression

- For MRR: In this case although the R^2 score is seen in the acceptable range, the RMSE value is found to be quite high.
- For TWR: Although the R^2 score in this instance is within acceptable bounds, the RMSE number is discovered to be extremely high.
- For Ra: The RMSE is discovered to be extremely high, despite the fact that the R^2 score in this case is within acceptable ranges.

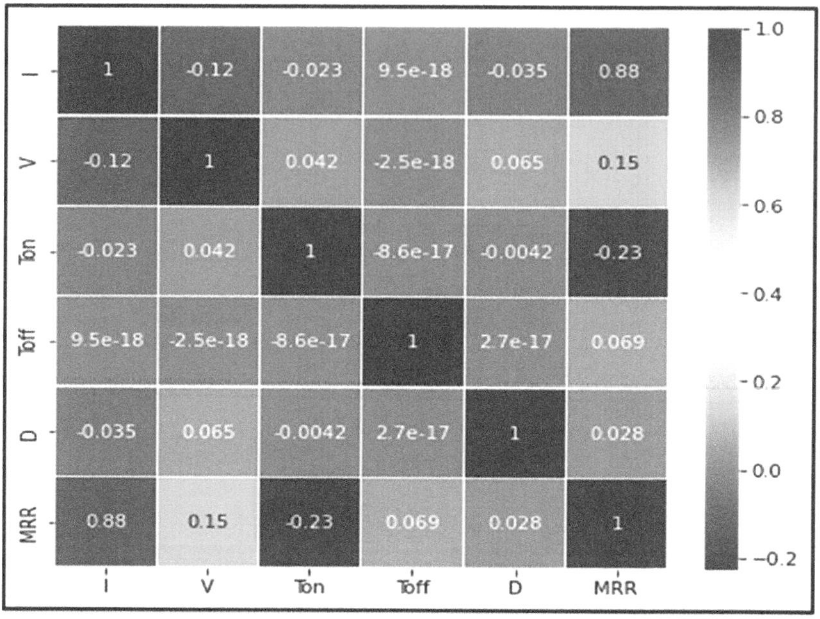

FIGURE 2.4
Correlation between input values and MRR.

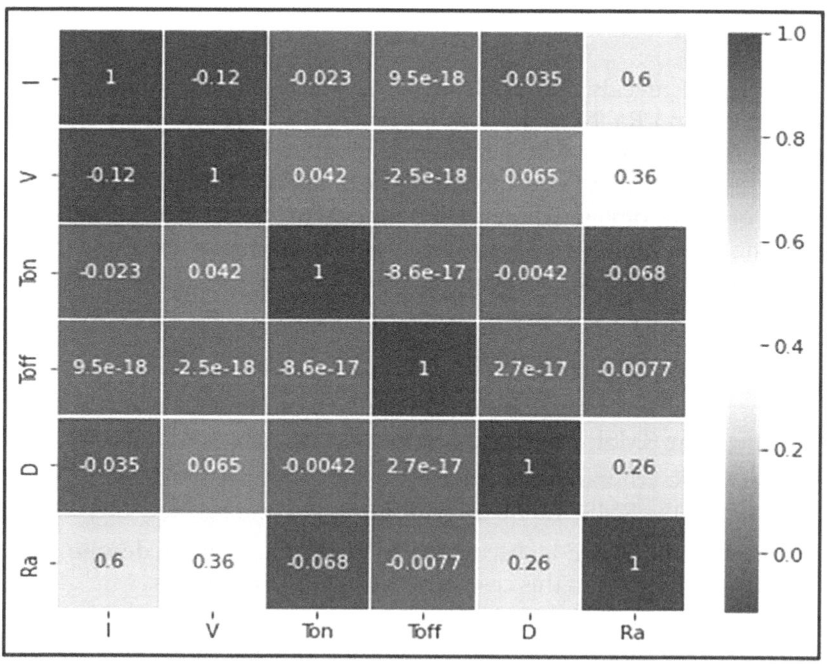

FIGURE 2.5
Correlation between input values and Ra.

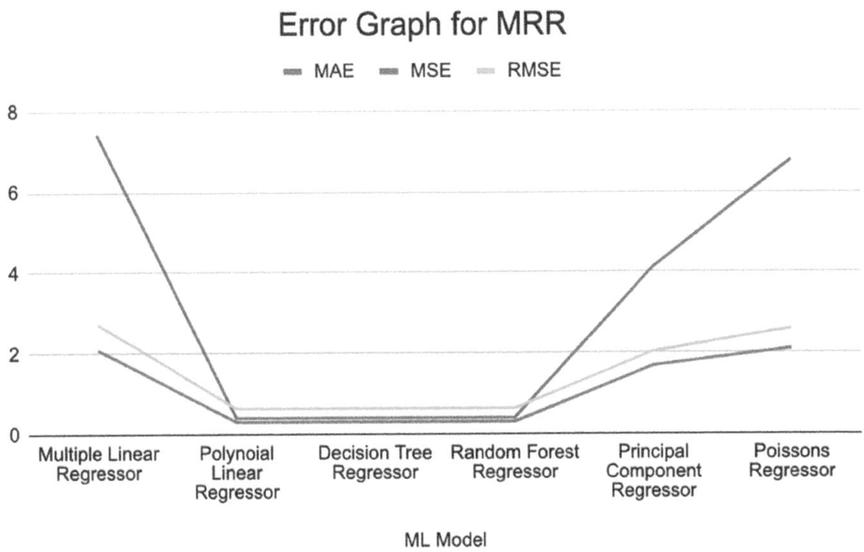

FIGURE 2.6
Comparative error scores for MRR.

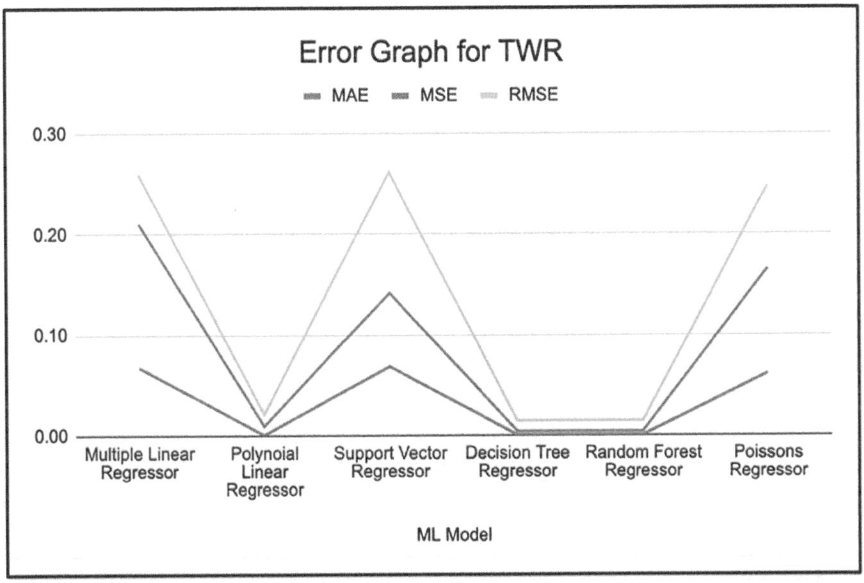

FIGURE 2.7
Comparative error scores for TWR.

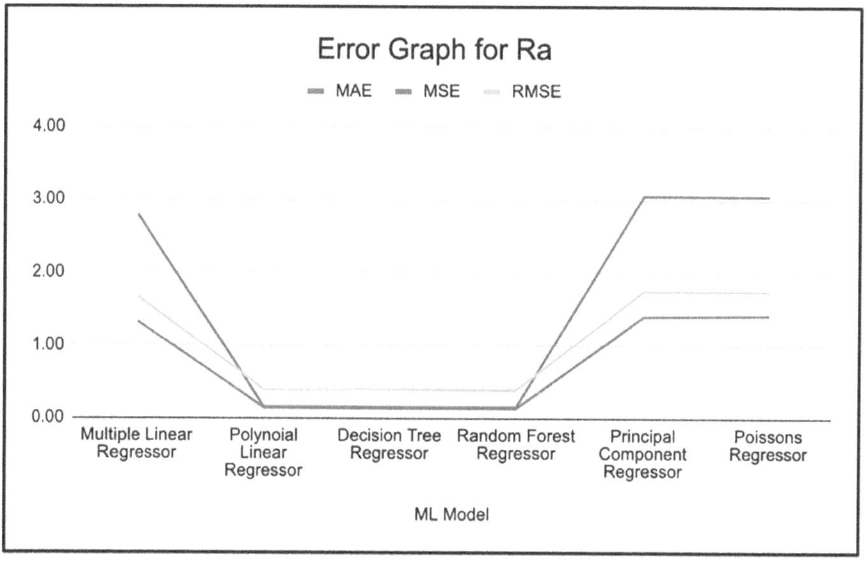

FIGURE 2.8
Comparative error scores for Ra.

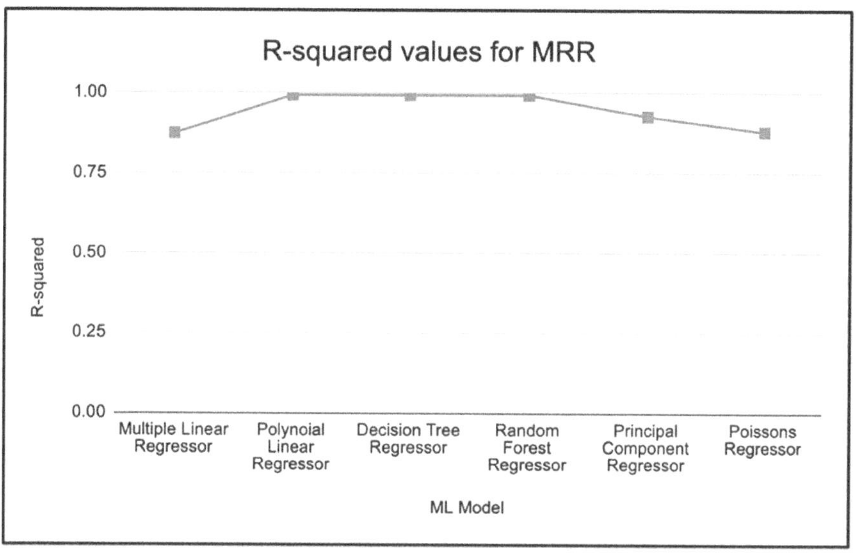

FIGURE 2.9
R-squared values of regression model – MRR.

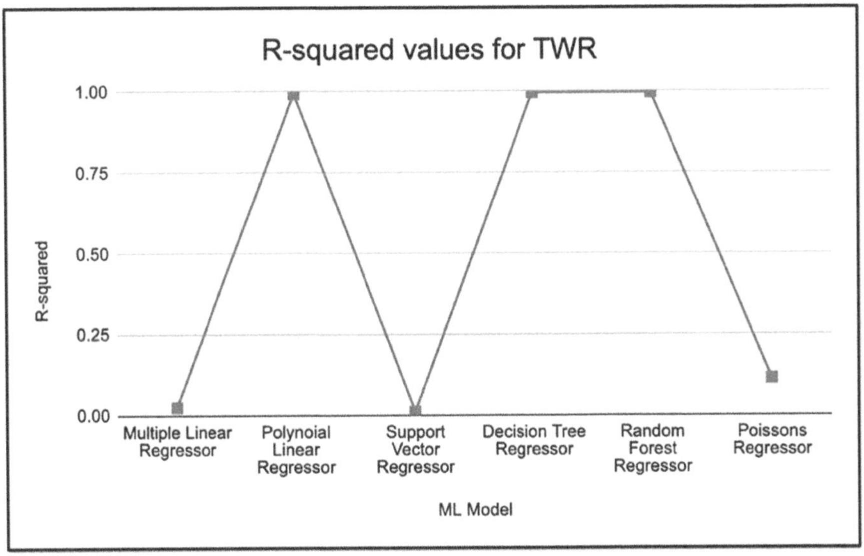

FIGURE 2.10
R-squared values of regression model – TWR.

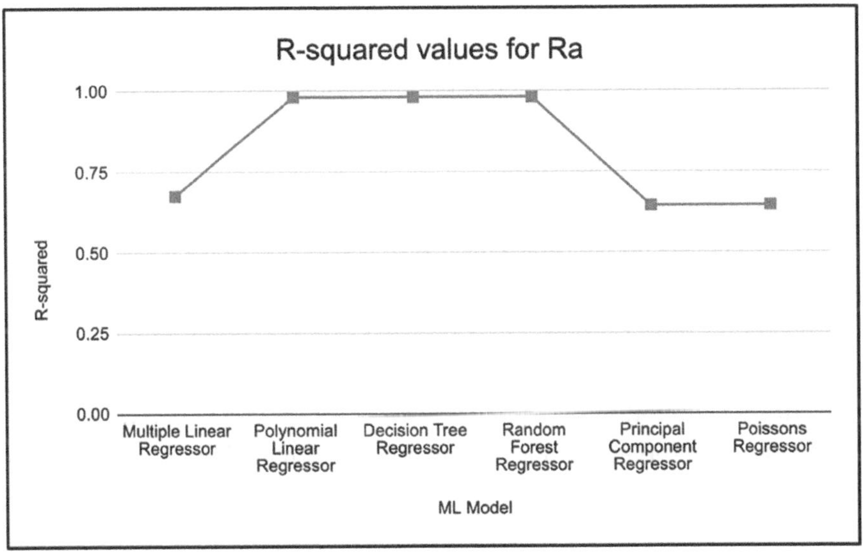

FIGURE 2.11
R-squared values of regression model – Ra.

2.3.2 Poisson's Regression

- For MRR: From the observed results, it is inferred that there is high variation and a high confidence margin, even though the model yields acceptable R^2 values. The very high error values that were found serve as additional evidence for this.
- For TWR: Even if the model produces acceptable R^2 values, it can be deduced from the observed results that there exist a large degree of variation and a high degree of confidence margin. The extremely high error values that were discovered provide further support for this.
- For Ra: The graph demonstrates that, despite the model producing acceptable error scores like MSE and MAE, there exist a substantial fluctuation and a high confidence margin. The light blue part denotes this margin of safety. This is further supported by the incredibly low R^2 values that were discovered.

2.3.3 Polynomial Linear Regression

- For MRR: This model's performance is deemed satisfactory due to its high R^2 score and low errors, particularly the RMSE.
- For TWR: This model is considered to be reliable because it has a high R^2 score and low errors, particularly the RMSE.
- For Ra: The high R^2 score and low inaccuracies, particularly the RMSE, indicate that the model is reliable and robust.

2.3.4 Support Vector Regression

- For MRR: The support vector machine (SVM) regression model does not give acceptable values for MRR.
- For TWR: Figure 2.7 shows that there is a significant variation and a high confidence margin, even though the model produces acceptable error scores like MSE and MAE. This confidence margin is indicated by the pale blue section. The extremely low R^2 values that were obtained further support this.
- For Ra: Acceptable values for Ra are not provided by the SVM regression model (Sawant et al., 2024).

2.3.5 Decision Tree Regression

By looking at the MRR reports, it is clear that the actual results do not have a huge gap with predicted results. A good R^2 and low RMSE make this kind of model preferred. Results of the TWR experiments and study of the errors in them demonstrate that this decision tree is a correct model to predict the

TWR. The results show that actual and estimated data from the testing process vary very little. It is a preferred model since it gives a large R^2 score and a small value for RMSE. The regressor was applied to Ra's dataset, and then graphs were drawn to check the predicted testing values with the expected results. It is inferred from the graph and the errors that are displayed that the decision tree is a well-suited approach for Ra prediction. From the graph, you can see how close the testing data's actual values and projections are. This model has a high R^2 and a low RMSE, so this is just ideal.

2.3.6 Random Forest Regression

Looking at MR, TWR, and Ra, the values in the graph for decision tree and random forest look the same, and the error values change very little compared to the algorithms. With these values, we can see that random forest regression beats decision tree regression (Jatti et al., 2024).

2.4 Conclusion

This work demonstrates a process that has potential application to materials and can be used to analyze and increase the sustainability of an EN19 machining process through the use of AI. During the investigation, we used AI technology to check critical factors related to gap current, time between pulses, voltage applied during a pulse, time when the pulse is applied, diameter of the tool, and environmental and sustainability impacts. Advanced correlation heat maps were used to reveal a list of key factors to consider in planning for sustainability and machining outcomes. Maximum sensitivity of surface quality sustainability indicators appeared to surface quality sustainability indicators for gap current, tool diameter, and gap voltage, while tool wear sustainability metrics were primarily controlled by pulse-on time and gap current combinations. A key process sustainability indicator, material removal efficiency, was found to be strongly affected by the synergistic effects between gap current and pulse-on time and gap voltage parameters. To predict surface quality, tool wear sustainability, and material removal efficiency, six machine learning regression models – linear, polynomial, principal component, Poisson's regression, random forest, and decision tree regression – were evaluated in the research. Random forest and decision tree models had better predictive power ($R^2 = 0.99$) and showed the promise that AI could be used to optimize sustainable manufacturing processes. These results provide a stable basis for implementing AI-driven solutions in the context of the sustainable manufacturing of composites and in nonconducting material applications.

References

Bhiradi, I., Raju, L., & Hiremath, S. S. (2020). Adaptive neuro-fuzzy inference system (ANFIS): Modelling, analysis, and optimisation of process parameters in the micro-EDM process. *Advances in Materials and Processing Technologies*, 6(1), 133–145.

Gaikwad, V., & Jatti, V. S. (2018). Optimization of material removal rate during electrical discharge machining of cryo-treated NiTi alloys using Taguchi's method. *Journal of King Saud University-Engineering Sciences*, 30(3), 266–272.

Ishfaq, K., Asad, M., Ashraf, W. M., Sana, M., Anwar, S., Zhang, W., & Dua, V. (2024). Towards artificial intelligence empowered performance enhancement of EDM process using nano-graphene mixed bio-dielectric supporting the carbon neutrality and sustainable development. *Journal of Cleaner Production*, 457, 142482.

Jain, S., & Parashar, V. (2021). Critical review on the impact of EDM process on biomedical materials. *Materials and Manufacturing Processes*, 36(15), 1701–1724.

Jatti, V. S., Krishnan, R. M., Saiyathibrahim, A., Preethi, V., Kumar, A., Sharma, S., & Lozanovic, J. (2024). Predicting specific wear rate of laser powder bed fusion AlSi10Mg parts at elevated temperatures using machine learning regression algorithm: Unveiling of microstructural morphology analysis. *Journal of Materials Research and Technology*, 33, 3684–3695.

Kamenskikh, A. A., Muratov, K. R., Shlykov, E. S., Sidhu, S. S., Mahajan, A., Kuznetsova, Y. S., & Ablyaz, T. R. (2023). Recent trends and developments in the electrical discharge machining industry: A review. *Journal of Manufacturing and Materials Processing*, 7(6), 204.

Kaur, R., Kumar, R., & Aggarwal, H. (2025). Systematic review of artificial intelligence, machine learning, and deep learning in machining operations: Advancements, challenges, and future directions. *Archives of Computational Methods in Engineering*, 1–54. https://doi.org/10.1007/s11831-025-10290-z

Pradhan, M. K., & Biswas, C. K. (2011). Multi-response optimisation of EDM of AISI D2 tool steel using response surface methodology. *International Journal of Machining and Machinability of Materials*, 9(1–2), 66–85.

Qudeiri, J. E. A., Zaiout, A., Mourad, A. H. I., Abidi, M. H., & Elkaseer, A. (2020). Principles and characteristics of different EDM processes in machining tool and die steels. *Applied Sciences*, 10(6), 2082.

Ranganathan, R., Saiyathibrahim, A., Velu, R., Jatti, V. S., Mohan, D. G., & Vijayakumar, P. (2025). Achieving multi-response optimization of control parameters for Wire-EDM on additive manufactured AlSi10Mg alloy using Taguchi-grey relational theory. *Engineering Research Express*, 7(1), 015404.

Sana, M., Asad, M., Farooq, M. U., Anwar, S., & Talha, M. (2024). Machine learning for multi-dimensional performance optimization and predictive modelling of nanopowder-mixed electric discharge machining (EDM). *The International Journal of Advanced Manufacturing Technology*, 130(11), 5641–5664.

Sawant, D. A., Jatti, V. S., Vibhute, A., Saiyathibrahim, A., Murali Krishnan, R., Bembde, S., & Balaji, K. (2024). Prediction of burn rate of ammonium perchlorate–hydroxyl-terminated polybutadiene composite solid propellant using supervised regression machine learning algorithms. *Aerospace Systems*, 1–9.

Sharma, D., Bhowmick, A., & Goyal, A. (2022). Enhancing EDM performance characteristics of Inconel 625 superalloy using response surface methodology and ANFIS integrated approach. *CIRP Journal of Manufacturing Science and Technology*, 37, 155–173.

Sharma, A., Kumar, V., Babbar, A., Dhawan, V., Kotecha, K., & Prakash, C. (2021). Experimental investigation and optimization of electric discharge machining process parameters using grey-fuzzy-based hybrid techniques. *Materials*, 14(19), 5820.

Singh, R., Singh, R. P., & Trehan, R. (2022). Machine learning algorithms based advanced optimization of EDM parameters: An experimental investigation into shape memory alloys. *Sensors International*, 3, 100179.

Surekha, B., Sree Lakshmi, T., Jena, H., & Samal, P. (2021). Response surface modelling and application of fuzzy grey relational analysis to optimise the multi response characteristics of EN-19 machined using powder mixed EDM. *Australian Journal of Mechanical Engineering*, 19(1), 19–29.

Upadhyay, C., Datta, S., Masanta, M., & Mahapatra, S. S. (2017). An experimental investigation emphasizing surface characteristics of electro-discharge-machined Inconel 601. *Journal of the Brazilian Society of Mechanical Sciences and Engineering*, 39(8), 3051–3066.

3

Implementation of AI and Industry 4.0 in Additive Manufacturing Processes

Swarn Singh, Harvinder Singh, Ankit Sharma, Ujval Mutyala, and Muralimohan Cheepu

3.1 Introduction

Artificial intelligence (AI) and Industry 4.0 technologies converge with additive manufacturing (AM) to change paradigms for production, enabling levels of precision, efficiency, and customization that are otherwise impossible. Integrating AI-driven analytics, real-time process monitoring, and interconnected cyber-physical systems enables manufacturers to optimize design workflows, predict equipment failure, and produce superior products. Industry 4.0, characterized by the digital architecture that enables Internet of Things (IoT)-connectivity and data-driven decision-making, augurs for AM's layer-by-layer manufacturing with reduced material wastage while speeding up prototyping. Therefore, mass customization will become feasible. This report highlights the transformational power of these technologies along design optimization, predictive maintenance, and quality assurance for sustainable production, with several case studies by the leading adopters of industrial applications. AM, also known as three-dimensional (3D) printing, is creating objects by sequentially laying down layers of material within the confines of a digital model 13. Cutting material waste, lacking the complexity of geometries that could be realized by conventional methods, adds and manufactures additive. This computer-aided design (CAD) has a facility to digitally represent blueprints, which could then be executed with the printer using polymers, metals, or composite materials1. This flexibility makes AM very important into the modern manufacturing space, especially in aerospace, healthcare sectors, and automobiles that need customization and prototyping in a very short time (Hose *et al.*, 2023).

The relationship between AI and AM processes is perhaps one of the most powerful drivers for the paradigm shift in production efficiency, resulting

 DOI: 10.1201/9781003506492-3

in factories achieving levels of precision, speed, and resource optimization that were previously unattainable. AI-based design algorithms into effective action are combined with the latest procedures in real-time process monitoring, predictive maintenance frameworks, and automated quality assurance systems so that AM technologies could begin to create impact on the overall operational results while minimizing material waste and energy consumption. This report examines AI's multifaceted role in enhancing AM efficiency, wherein empirical data from industrial applications and future technological advancements is provided (Mohapatra, 2024).

3.2 Intelligent Manufacturing

AM largely alters the use of materials in production systems. The systematic reduction of generation waste occurs by the way fabrication is done layer by layer. Eliminating all forms of subtractive processes, optimizing material use with AI-driven design, and implementing closed-loop material recycling in additive manufacturing lead to waste reductions of 70 to 90% compared to conventional manufacturing methods (Kumar *et al.*, 2024; Di & Yang, 2022). This report analyzes six core mechanisms through which AM minimizes waste, supported by empirical data from aerospace, automotive, and medical industries (Taher & Basher, 2024) (Figure 3.1).

3.2.1 Inherent Material Efficiency of Layer-by-Layer Fabrication

3.2.1.1 Elimination of Subtractive Waste

Traditional subtractive manufacturing methods, such as Computer Numerical Control (CNC) machining, discard 40–95% of raw materials as chips, shavings, or offcuts during shaping operations. In contrast, AM deposits material only where needed, reducing scrap generation to <5% for most polymer and metal processes (Bhaskaran *et al.*, 2024). For example, Airbus reduced titanium waste by 75% when switching from machining to laser powder bed fusion (LPBF) for A350 XWB bracket production.

3.2.1.2 Precision Material Deposition

Modern AM systems employ AI-controlled nozzles or lasers to deposit materials with ±0.1 mm accuracy, minimizing overspray and dimensional errors (Brueckner *et al.*, 2017). HP's Multi Jet Fusion technology achieves 99.8% material utilization rates in polymer part production through voxel-level control (Svetlizky *et al.*, 2021; Kanishka & Acherjee, 2023).

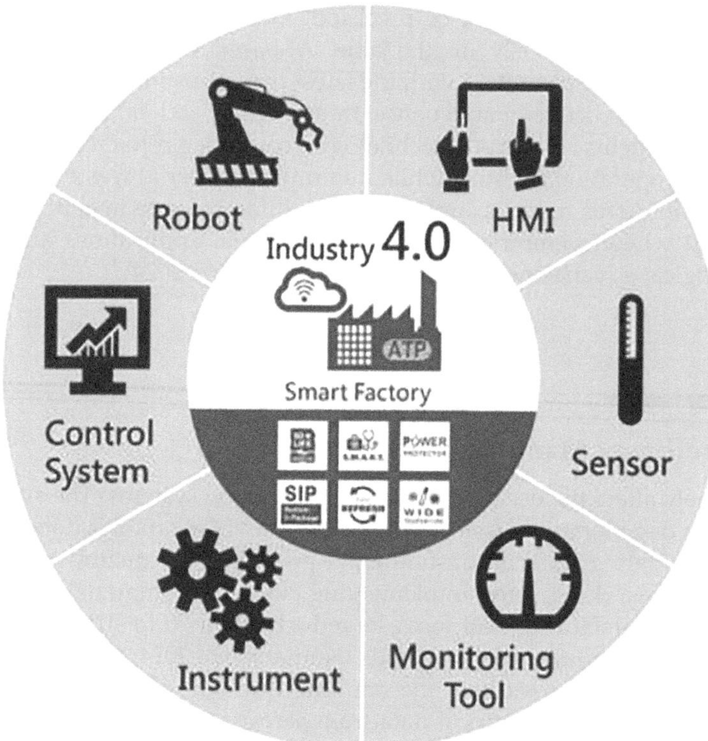

FIGURE 3.1
Components of intelligent manufacturing.

3.2.2 Design Optimization Strategies for Minimal Material Use

3.2.2.1 Topology Optimization and Generative Design

AI-driven topology optimization tools like Autodesk's Fusion eliminate structurally redundant material while maintaining load-bearing capacity (10). General Electric's LEAP engine bracket redesign using these methods reduced weight by 84% and material consumption by 63%, saving $31 million annually in fuel costs (Vlah *et al.*, 2020).

3.2.2.2 Support Structure Minimization

Support structures account for 15–40% of material use in AM. Innovations like cellular support systems and self-supporting lattice designs have reduced support requirements by 45% in selective laser melting (SLM) applications (Jiang *et al.*, 2018; Gan & Wong, 2016). Materialise's Bluesint PA 12 technology enables powder bed fusion without supports for certain geometries, eliminating 100% of support-related waste.

3.2.3 Closed-Loop Material Recycling Systems

3.2.3.1 Metal Powder Recovery and Reconditioning

Metal AM processes leave 90–95% of unused powder after printing. Advanced recovery systems from companies like EOS and GE Additive sieve, degas, and blend residual powders to restore flow ability and particle size distribution. X-ray diffraction analysis confirms that IN718 nickel alloy powder can undergo 15 reuse cycles before oxidation necessitates recycling (Gan & Wong, 2016; Yi *et al.*, 2021).

3.2.3.2 Polymer Filament Regeneration

Fused deposition modeling printing waste is crushed, melted, and extruded into new filament through devices like INTAMSYS' FUNMAT PRO 610, achieving 97% recycling efficiency (Shiferaw & Gebremedhen, 2022). Formfutura's ReForm rPETG filament contains 100% post-industrial recycled content, diverting 12 tons/month of plastic from landfills.

3.2.4 On-Demand Production Models

3.2.4.1 Digital Warehousing and Localized Manufacturing

AM Efficiency's distributed production network maintains CAD files in digital warehouses, printing components at regional hubs only when orders are placed. This approach eliminated 83% of inventory waste for a medical device client by preventing overproduction (Demartini *et al.*, 2021).

3.2.4.2 Mass Customization without Tooling Changes

Dental aligner manufacturers like Align Technology produce 500,000 unique designs daily using Carbon's DLS printers, avoiding the material waste associated with retooling injection molds for each patient (Da Silveira *et al.*, 2001).

3.3 Energy and Resource Efficiency Synergies

3.3.1 Reduced Energy Intensity

AM consumes 25–30% less energy per kilogram of output compared to machining aluminum alloys, as layer-wise fusion requires less peak power than bulk material removal (Duflou *et al.*, 2012; van den Berg *et al.*, 2016). Fraunhofer Institute's adaptive laser control systems further reduce energy use by 22% in LPBF processes.

3.3.2 Light Weighting Effects

BMW's 3D-printed brake calipers weigh 40% less than cast equivalents, reducing vehicle energy consumption by 0.2 kWh/km. Scaled across 100,000 electric vehicles, this yields annual CO_2 savings equivalent to 12,000 tons of avoided aluminum production (Veza *et al.*, 2023).

3.4 Circular Economy Integration

3.4.1 Component Repair Instead of Replacement

Siemens Energy repairs gas turbine blades via directed energy deposition (DED), using 89% less material than full replacements (Walmsley *et al.*, 2019). This approach extended blade service life by 15,000 hours in a Malaysian power plant application.

3.4.2 End-of-Life Material Recovery

Stratasys' Infinite-Build 3D Demonstrator incorporates dissolvable support materials that enable 100% part disassembly. Post-use ABS components are pyrolyzed into feedstock for new printers, achieving 92% material circularity (Chatziparaskeva *et al.*, 2022; Pratap *et al.*, 2024).

3.4.3 Waste Reduction Metrics across Sectors

Aerospace: 67–83% material waste reduction in titanium components.

Automotive: 55% lower polymer scrap rates versus injection molding.

Healthcare: 90% reduction in surgical guide waste through patient-specific printing.

3.4.4 Emerging Technologies

1. AI-Predictive Material Allocation: Algorithms forecast optimal material quantities per build, reducing over-powdering by 30% (Ma *et al.*, 2024).
2. Multi-Material Recycling: Hybrid systems separating PLA/PETG blends achieve 85% purity for closed-loop reuse.
3. 4D Printing Self-Recycling: Thermally responsive polymers that self-disassemble into reusable pellets after service life.

AM establishes a new waste hierarchy in industrial production, systematically addressing material inefficiencies through:

1. AI-optimized designs eliminating redundant material.
2. Layer-wise precision minimizing process waste.
3. Closed-loop powder recovery systems.
4. Advanced filament regeneration technologies.

With global AM adoption projected to prevent 130 million tons of manufacturing waste by 2030, the technology represents a cornerstone of circular economy transitions (Dou *et al.*, 2023). Future advancements in machine learning (ML)-driven process control and standardized material recycling protocols will further amplify AM's waste mitigation potential across supply chains (Soares *et al.*, 2022).

3.4.5 Industry 4.0: The Digital Backbone of Smart Factories

Industry 4.0 represents the Fourth Industrial Revolution, marked by the fusion of physical production systems with digital technologies such as IoT, cloud computing, and big data analytics (Nguyen *et al.*, 2019). Smart factories equipped with embedded sensors and robotics collect real-time data, enabling predictive maintenance, self-optimizing processes, and seamless integration across supply chains (Sahoo & Lo, 2022). This digital transformation allows manufacturers to transition from reactive to proactive operations, leveraging data to anticipate market demands and mitigate disruptions (Soori *et al.*, 2023; Gao *et al.*, 2020).

AI encompasses ML, computer vision, and cognitive computing, enabling machines to analyze data, learn patterns, and make autonomous decisions. In manufacturing, AI algorithms optimize production parameters, detect defects, and predict equipment failures, fostering a shift from human-dependent oversight to automated, intelligent systems (Rashid & Kausik, 2024). When applied to AM, AI enhances design complexity, process reliability, and resource efficiency, creating a symbiotic relationship with Industry 4.0's interconnected infrastructure (Anantrasirichai & Bull, 2022).

3.4.6 Digital Transformation of AM through Industry 4.0

Industry 4.0's IoT framework facilitates bidirectional communication between AM machines, enterprise resource planning systems, and supply chain networks. Sensors embedded in 3D printers monitor variables such as temperature, pressure, and material flow, transmitting data to centralized platforms for analysis (Munirathinam, 2020). This real-time visibility enables operators to adjust printing parameters dynamically, ensuring consistent output quality despite fluctuating environmental conditions. For example, GE Additive employs AI-powered systems to modulate laser power and scanning speed during metal 3D printing, compensating for thermal distortions and minimizing defects (Tuli *et al.*, 2024). Digital twins—virtual replicas of physical assets—play a pivotal role in optimizing AM processes. By simulating printing operations, manufacturers can identify potential failures, test design modifications, and validate production strategies before physical execution. Siemens utilizes digital twins to predict wear-and-tear in 3D printers, scheduling maintenance interventions that prevent unplanned downtime. This integration of virtual and physical realms reduces prototyping cycles and enhances operational agility, aligning with Industry 4.0's emphasis on data-driven innovation.

3.4.7 AI-Driven Automation in AM

AI effects a revolution in design phases by generating lightweight, structurally orthodox components to improve the performance of materials. According to Autodesk, their generative designing tools exploit ML in order to draw thousands of possibilities for designs, sifting through configurations that fulfill load-bearing while minimizing the use of material. In such cases, a weight gain saves much fuel cost, which makes it groundbreaking in aerospace. By taking away the bounds of human thinking, AI directs toward geometries that include strength, durability, and resource conservation (Soori *et al.*, 2025). Such abnormalities as misalignment of understood layers or porosity are detected by AI systems analyzing the data from sensors during printing. At HP, the Multi Jet Fusion technology uses computer vision to scan every layer in real time and, if required, automatically rectify deviations before they would propagate. In this sense, adaptive controls cause first-pass success rates, mostly above 95%, involving reduced after-processing and scrap cost. Moreover, ML models trained from historical data forecast optimum printing parameters for new materials, allowing faster qualification and applicability of AM.

Manufacturers end up losing about $50 billion a year on unplanned equipment downtimes, which might have been tackled through the predictive maintenance of AI. For example, ML algorithms can determine within weeks whether problems might arise in a particular component by analyzing the 3D printer's vibration, thermals, and acoustics data. Siemens' AI platforms, for example, relate sensor readings with maintenance records to spot early signs of degradation in motors or nozzle clogs. If a repair is made before breakdown occurs, the equipment can be made to last longer because the equipment does not stop production while repairs would have normally been occurring, which is very important for high-volume AM applications such as medical implants (36). AI-powered vision systems achieve submicron accuracy, and their outputs are not subject to human errors that are commonly found in traditional labor-intensive quality inspections. Affected by deep learning models that use defect database, such anomalies as cracks, warping, or incomplete fusion have a 99% precision of classification. For example, turbine blades are scanned very accurately by BMW after printing with the help of a comparison with a CAD model derived from 3D scans. Automated feedback loops suitably modify the printing parameters to any deviations in a given printing process, thus ensuring adherence to stringent aerospace and medical standards (Martin *et al.*, 2005).

Generative design algorithms run by AI revolutionize the entire conceptualization phase of AM components. The Autodesk Fusion platform embodies this radical transformation so that all engineers can input a performance criterion capacity to bear load, thermal resistance, weight limits, and many more. The system will then generate thousands of permutations for designs, whereas each can be assessed through physics-based simulations to identify optimal geometries. Thus, as an illustration, lattice formations such as these would reduce by around 30–40% material usage without compromising strong structures, thereby saving costs yet more importantly promoting sustainability.

3.5 Real-Time Process Monitoring and Adaptive Control Systems

Modern AM systems integrate IoT sensors and high-resolution cameras to collect real-time data on temperature gradients, layer adhesion quality, and nozzle alignment. AI algorithms analyze these data streams at millisecond intervals, dynamically adjusting laser power, printing speed, and material deposition rates to maintain optimal conditions (Rajak *et al.*, 2023). HP's Multi Jet Fusion technology employs such systems to achieve first-pass success rates exceeding 95%, reducing scrap rates and postprocessing requirements. The implementation of edge computing architectures enables localized data processing, minimizing latency during critical printing phases. For example, temperature fluctuations in metal AM processes can be corrected within 0.5 seconds using AI controllers, preventing warping defects that traditionally required scrapping entire components (Gouiza *et al.*, 2024) (Figure 3.2).

3.6 AM with IOT For Real-Time Monitoring

The integration of AM with IoT technologies has revolutionized production processes by enabling real-time monitoring, predictive maintenance,

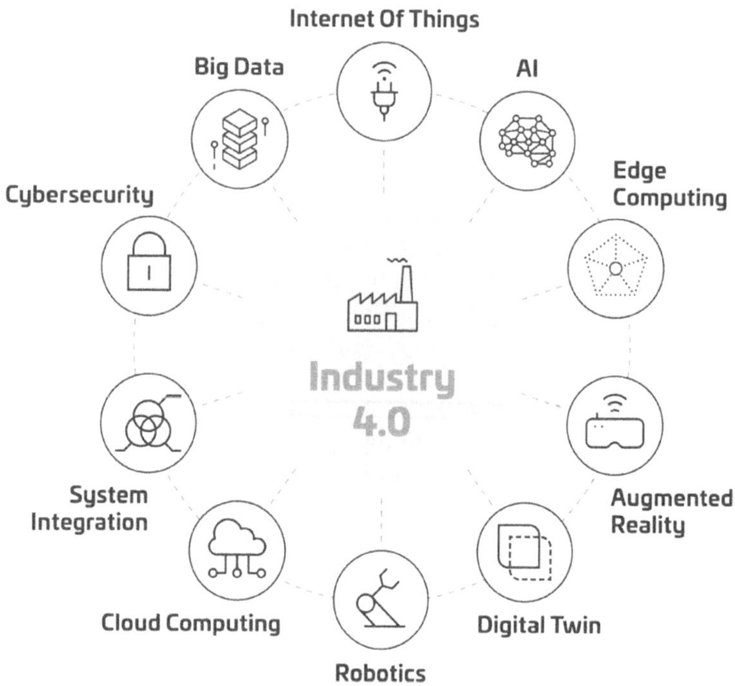

FIGURE 3.2
Role of IoT in real-time process monitoring.

and enhanced quality control. IoT sensors embedded in AM systems collect critical data on temperature, humidity, print quality, and machine performance, allowing for immediate adjustments to optimize production efficiency and reduce material waste (Morchid *et al.*, 2025). This report outlines the key mechanisms through which IoT enhances AM operations, supported by industrial applications and technological advancements (Dhatterwal *et al.*, 2025).

3.6.1 Real-Time Data Collection and Monitoring

IoT sensors installed on AM machines continuously monitor parameters such as temperature, humidity, and print quality. This real-time data collection allows manufacturers to identify potential issues early, reducing the likelihood of production errors and material waste (Sani *et al.*, 2024). For instance, temperature fluctuations during metal AM processes can be corrected within seconds using IoT-driven feedback loops, preventing defects like warping or porosity (Soori *et al.*, 2025).

3.6.2 Predictive Maintenance and Reduced Downtime

Predictive maintenance is an important benefit of IoT integration in AM. By analyzing sensor data, manufacturers can detect anomalies in machine performance before they cause breakdowns, minimizing unplanned downtime and improving overall efficiency (Lu *et al.*, 2024). This proactive approach extends equipment lifespan, reduces maintenance costs, and ensures continuous production schedules for high-value components like aerospace parts (MajidiParast *et al.*, 2025).

3.6.3 Enhanced Quality Control and Customization

IoT sensors provide detailed insights into the production process, enabling precise control over part quality. Real-time monitoring allows for early detection of defects, which can be corrected immediately to maintain high-quality standards (Pagliari *et al.*, 2025). Additionally, IoT facilitates mass customization by tracking production in real time, enabling adjustments to meet specific customer requirements (Korzenowski *et al.*, 2013).

3.6.4 Supply Chain Integration and Inventory Management

IoT integration enables seamless data sharing across supply chains, enhancing inventory management and resource allocation. Real-time visibility into production status and inventory levels helps managers optimize material ordering and reduce stock-outs, streamlining the entire product development process (Puttemans *et al.*, 2025). This connectivity supports just-in-time manufacturing, allowing companies to respond quickly to market demands while minimizing waste.

3.6.5 Remote Monitoring and Control

IoT technologies, combined with high-speed connectivity solutions like 5G and WiFi-6, enable remote monitoring and control of AM systems. Operators can access real-time data and adjust machine parameters from anywhere, ensuring safety and efficiency while reducing the need for on-site personnel. This capability is particularly valuable in distributed manufacturing networks, where real-time oversight is crucial for maintaining quality and productivity across multiple locations (Shukla *et al.*, 2025).

As AM continues to integrate with IoT, future advancements will likely involve deeper AI integration for predictive analytics and automated decision-making. Challenges include ensuring data interoperability across different IoT platforms and maintaining cybersecurity in increasingly connected manufacturing environments. However, the benefits of IoT-enhanced AM such as improved efficiency, reduced waste, and enhanced customization position this technology at the forefront of Industry 4.0's smart manufacturing initiatives (Ali *et al.*, 2025).

3.7 Sustainability and Resource Efficiency In AI-Enhanced AM

3.7.1 Material Optimization and Waste Reduction

AM's additive nature inherently reduces material waste compared to subtractive methods, but AI further enhances sustainability. ML algorithms optimize support structures and nesting layouts, minimizing raw material consumption by up to 40%. For example, AI-driven software at Airbus reorients aircraft components during printing to reduce titanium usage without compromising structural integrity (Martins *et al.*, 2025). Additionally, generative designs often incorporate lattice structures that use 70% less material than solid counterparts, aligning with circular economy principles.

3.7.2 Energy Efficiency and Carbon Footprint Mitigation

AI optimizes energy consumption by scheduling printing jobs during off-peak hours and regulating power usage based on real-time grid data. A study by Fraunhofer Institute demonstrated that AI-controlled 3D printers reduce energy consumption by 25% through dynamic adjustments to laser intensity and cooling systems. Furthermore, localized AM production eliminates transportation-related emissions, as digital files can be transmitted globally for on-site printing. This decentralization supports sustainable manufacturing practices while reducing logistics costs (Martins *et al.*, 2025).

AI-driven predictive maintenance systems analyze vibration patterns, thermal signatures, and acoustic emissions from AM printers to forecast component failures with 92% accuracy. By correlating historical maintenance records with real-time sensor data, these models identify early warning signs such as bearing wear or laser diode degradation (Aeddula *et al.*, 2024). Siemens' implementations demonstrate a 40–50% reduction in unplanned downtime through such systems, ensuring continuous production schedules for high-value components like turbine blades. The financial impact is substantial: Manufacturers report a 20–40% decrease in maintenance costs by transitioning from calendar-based to condition-based servicing. This optimization extends equipment lifespan by 25–30%, delaying capital expenditures for machine replacements.

3.8 Industrial Applications of AI and Industry 4.0 in AM

Transformative AI has penetrated diverse industries to revolutionize a very high degree of efficiency in operation, product, as well as customer engagement. The applications using AI such as ML, computer vision, and natural language processing could bring in transformational changes into predicting maintenance, automation, generative design, and supply chain management. This chapter presents all the big areas on the application of AI in industry, lending itself to the manufacturing industry as the prime cite of such transformations from human to AI.

3.8.1 Predictive Maintenance and Equipment Optimization

One of the pillars of modern manufacturing, predictive maintenance powered by AI allows companies to analyze machinery sensor data and predict possible failures before they occur. In Siemens' industrial applications, this has resulted in decreases in unplanned downtime by as much as 63% and maintenance costs by 20–40%. When the IoT sensors are merged with AI, the manufacturers can schedule maintenance during non-peak times, minimizing production disruptions (Bousdekis *et al.*, 2017).

3.8.2 Automation and Assembly Line Optimization

Efficiency in assembly lines is driven by AI automation through its analysis of past performance metrics and actual sensor data so as to establish better workflow and improved downtimes. For instance, one such incorporation of AI is to optimize assembly line operations for Volkswagen, enhancing the accuracy and speed of production while adapting themselves in accordance with the demand modification. AI also automates tasks that are repetitive, just like quality control inspections, which guarantees uniformity and accuracy in all measures done at the end of the production process (Zheng *et al.*, 2025).

3.8.3 Generative Design and Product Innovation

AI algorithms are at the heart of generative design, a transformative phenomenon in product development, wherein literally thousands of permutations of design get generated based on input parameters, such as material constraints and structural requirements. With the aid of AI, Airbus was able to improve the prediction of aerodynamic forces and moments from 1 hour to 30 milliseconds, thus enhancing the innovation capacity by an order of magnitude. The rapid design development and prototyping life cycles that this approach affords permit manufacturers to optimize designs with respect to performance, weight, and cost (Wang *et al.*, 2025).

3.8.4 Supply Chain Management and Logistics Optimization

Demand forecasting, inventory management, and logistics optimization are done using ML algorithms that promote supply chain resilience. For example, Walmart uses AI for demand forecasting, optimizing inventory levels while minimizing stock-outs, to ensure product availability and limit excess inventory. AI-powered routing algorithms further optimize delivery routes, enhancing customer satisfaction and reducing logistics costs (Qu, 2024).

3.8.5 Quality Assurance and Defect Detection

AI-powered quality control systems employ computer vision to analyze images or videos of products, accurately identifying defects and anomalies with higher precision than human inspectors. Foxconn, a leading electronics manufacturer, uses AI to detect flaws in electronic components, ensuring that products meet strict quality standards. This real-time monitoring enhances production efficiency and accuracy, allowing companies to produce high-quality goods at scale (Psarommatis, 2024).

As AI continues to evolve, future applications will likely involve deeper integration with Industry 4.0 technologies, such as IoT and digital twins, to create more interconnected and adaptive manufacturing ecosystems. Challenges include ensuring data interoperability across different AI platforms and addressing workforce displacement concerns. However, the benefits of AI in industry such as improved efficiency, reduced waste, and enhanced innovation position it as a critical component of modern industrial strategy.

AI's applications in industry, particularly in manufacturing, have transformed operational efficiency, product innovation, and customer engagement. By leveraging AI for predictive maintenance, automation, generative design, and supply chain optimization, companies achieve significant reductions in waste, energy consumption, and production costs while enhancing product quality and customization (Ma *et al.*, 2024). As AI technologies continue to advance, they will remain at the forefront of industrial innovation, driving Industry 4.0's smart manufacturing initiatives forward (Saha *et al.*, 2025).

GE Additive integrates AI with direct metal laser sintering (DMLS) to produce turbine blades for jet engines. ML algorithms analyze thermal imaging data to adjust laser parameters, preventing residual stress and improving fatigue resistance. Digital twins simulate airflow dynamics, ensuring optimal cooling channel geometries that enhance engine efficiency by 15%. This application underscores AI's role in advancing high-stake AM components.

Siemens Healthineers employs AI to customize orthopedic implants based on patient CT scans. Generative design algorithms create porous structures that promote bone integration, while real-time monitoring ensures biocompatibility and dimensional precision. Post-printing, AI systems validate implant surfaces against regulatory standards, reducing rejection rates by 30%. This patient-centric approach exemplifies AM's potential in personalized medicine.

BMW leverages AI-driven topology optimization to design lightweight brake calipers for electric vehicles. By iterating through thousands of design variations, engineers achieved a 40% weight reduction without sacrificing performance. IoT sensors in production lines monitor layer adhesion, automatically pausing printers if deviations exceed tolerance thresholds. This integration of AI and Industry 4.0 has shortened BMW's prototyping cycles from months to weeks.

3.9 Automated Quality Assurance and Defect Mitigation

Computer vision systems equipped with convolutional neural networks (CNNs) perform micron-level inspections of each printed layer, comparing 3D scan data against CAD models in real time. BMW's AM facilities utilize this technology to detect porosity defects in automotive components, achieving a 99.7% detection accuracy rate that surpasses human inspection capabilities. When deviations exceed tolerance thresholds, the system automatically pauses printing and suggests parameter adjustments, preventing the propagation of errors through subsequent layers. Postproduction quality checks benefit from AI's ability to analyze X-ray tomography data for internal voids or residual stresses. This capability reduces manual inspection labor by 70% while improving compliance with aerospace and medical device certification standards (Kehayias *et al.*, 2025).

3.9.1 Material and Energy Efficiency Optimization

ML algorithms optimize material usage through two primary mechanisms:

1. Support Structure Minimization: By analyzing stress distribution patterns, AI determines the minimal support framework required during printing, reducing material consumption by 15–25% in polymer-based AM.

2. Nesting Optimization: AI-driven nesting algorithms arrange multiple components within the build chamber to maximize space utilization, achieving 92–95% packing efficiency compared to manual layouts.

Energy consumption patterns are similarly optimized through AI scheduling systems that align high-power sintering phases with off-peak electricity rates. Fraunhofer Institute studies confirm a 22–27% reduction in energy costs through such dynamic load balancing.

3.10 Case Study

3.10.1 Autodesk's Generative Design in Automotive Manufacturing

An automotive manufacturer implementing Autodesk's generative design tools reduced brake caliper weight by 40% while maintaining equivalent performance characteristics. The AI-optimized design decreased material costs by $18,500 per 1,000 units and shortened the development cycle from 14 weeks to 9 days.

3.10.2 Predictive Maintenance in Aerospace AM Facilities

An aerospace parts producer integrated vibration analysis AI across 27 metal AM systems, achieving:

- 63% reduction in unplanned downtime.
- $2.7 million annual savings in maintenance costs.
- 19% increase in machine utilization rates.

3.11 Challenges and Future Directions

3.11.1 Technical and Operational Barriers

Despite its promise, AI–AM integration faces challenges such as data silos, algorithm bias, and high initial investment costs. Legacy machines lacking IoT connectivity require retrofitting, while proprietary software ecosystems hinder interoperability. Additionally, ML models trained on limited datasets may underperform in novel scenarios, necessitating continuous data enrichment.

3.11.2 Ethical and Workforce Implications

The shift toward autonomous AM systems raises concerns about job displacement, particularly in quality control and manual design roles. However, this transition also creates opportunities for upskilling workers in AI oversight and digital twin management. Ethical frameworks must address data privacy risks, ensuring that sensitive design files and production metrics remain secure.

3.11.3 Emerging Trends: Quantum Computing and Advanced Materials

Future advancements may see quantum computing accelerating generative design iterations, enabling real-time optimization of multi-material components. Similarly, AI-driven discovery of novel alloys and composites could expand AM's applicability to extreme environments. The integration of block-chain for supply chain transparency and AI-powered digital twins for cross-factory collaboration will further solidify AM's role in Industry 4.0 ecosystems.

3.12 Benefits of Integrating Industry 4.0 Technologies with AM

The fusion of Industry 4.0 technologies with AM has unlocked transformative advantages across production ecosystems, driving efficiency, customization, and sustainability. By embedding IoT-enabled sensors, AI-driven analytics, and digital twins into AM workflows, manufacturers achieve unprecedented operational agility, predictive capabilities, and resource optimization. This report synthesizes empirical evidence and industrial case studies to delineate the core benefits of this integration, emphasizing waste reduction, real-time process control, and intelligent supply chain management.

3.12.1 Enhanced Material Efficiency and Waste Reduction

Industry 4.0's data-centric approach synergizes with AM's additive nature to minimize material waste. Traditional subtractive manufacturing methods discard up to 90% of raw materials during machining, whereas AM builds components layer-by-layer, using only the required material. IoT sensors embedded in 3D printers monitor material flow and deposition rates, dynamically adjusting parameters to optimize consumption (Ullah *et al.*, 2024). For instance, AI algorithms reduce support structure requirements by 25–40% through stress distribution analysis, as demonstrated in Airbus's titanium

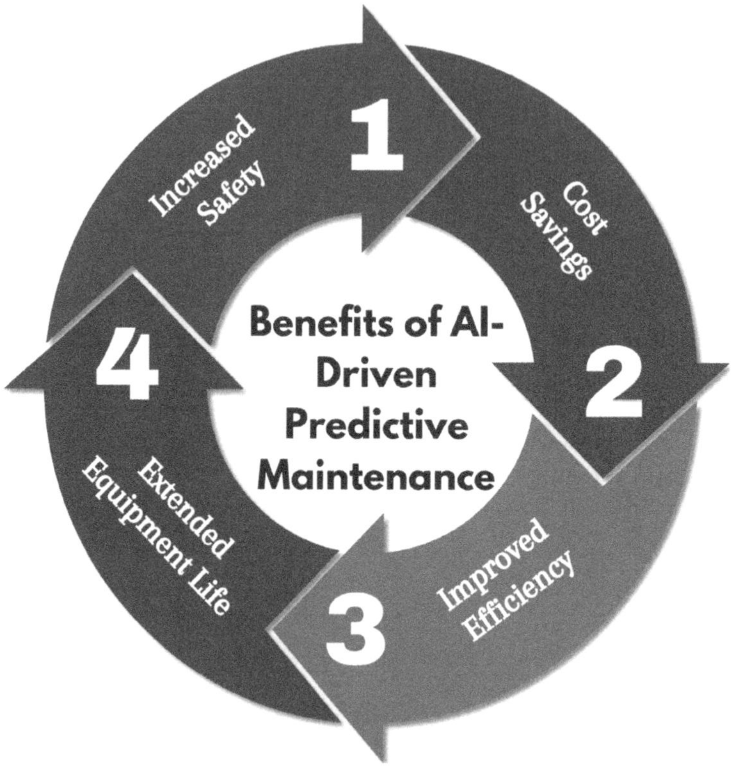

FIGURE 3.3
Benefits of AI-driven maintenance.

aerospace components. This precision not only lowers raw material costs but also aligns with circular economy principles by reducing scrap generation (Dias *et al.*, 2023) (Figure 3.3).

3.12.2 Accelerated Prototyping and Design Iteration

The integration of generative design tools and digital twins compresses product development cycles from weeks to days. AI-powered software like Autodesk's Fusion generates thousands of design permutations, evaluating structural integrity and thermal performance through physics-based simulations. Engineers can rapidly prototype and test components, iterating designs in response to real-time feedback. BMW's implementation of generative design for brake calipers reduced development time by 65%, achieving a 40% weight reduction without compromising performance. Digital twins further enhance this process by simulating manufacturing outcomes, enabling virtual validation before physical production.

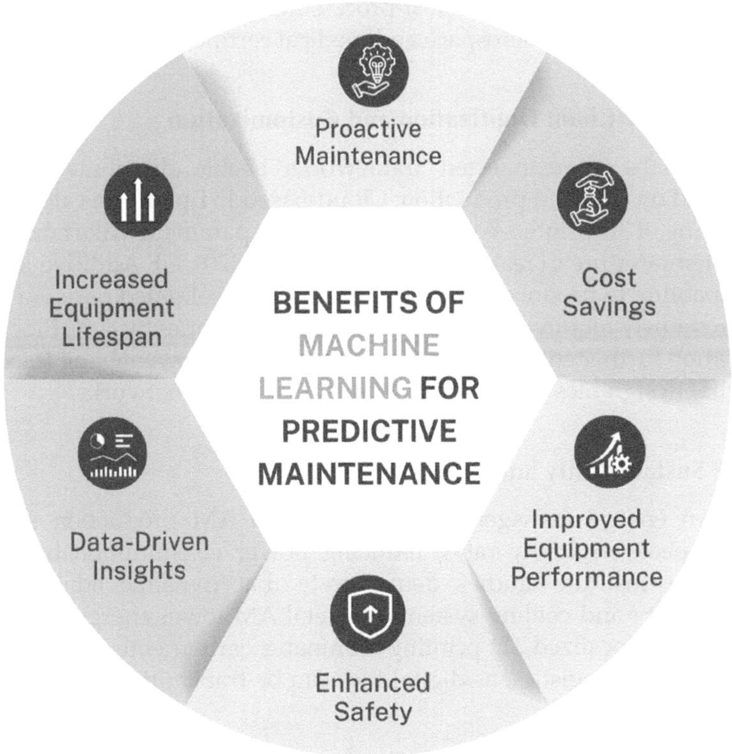

FIGURE 3.4
Benefits of machine learning for predictive maintenance.

3.12.3 Predictive Maintenance and Operational Continuity

Condition-based predictive maintenance, a cornerstone of Industry 4.0, mitigates unplanned downtime in AM systems. ML models analyze vibration, thermal, and acoustic data from IoT sensors to forecast equipment failures with 92% accuracy (Lettori *et al.*, 2020). For example, Siemens' AI platforms predict laser diode degradation in metal AM printers, enabling proactive servicing that reduces downtime by 63% in aerospace applications. This shift from reactive to predictive maintenance slashes annual maintenance costs by 20–40%, while extending equipment lifespan by 25–30% (Figure 3.4).

3.12.4 Real-Time Process Monitoring and Quality Assurance

IoT-enabled sensors and computer vision systems ensure micron-level precision during printing. HP's Multi Jet Fusion technology employs high-resolution cameras to inspect each layer, automatically correcting deviations in nozzle alignment or material deposition. BMW's AM facilities report 99.7% defect detection accuracy using CNNs to compare 3D scans with CAD models. Such systems reduce post-processing labor by 70% and ensure compliance with stringent aerospace and medical certification standards.

3.12.5 Supply Chain Digitization and Customization

Industry 4.0's interconnected frameworks enable distributed manufacturing and on-demand production. Cloud-based IoT platforms allow remote monitoring of 3D printers worldwide, adjusting parameters like nozzle speed or bed temperature in real time (da Silva *et al.*, 2020). GE Additive leverages this capability to produce fuel nozzles with 21 consolidated parts—a process that previously required assembly—reducing lead times by 80%. Block-chain integration further enhances supply chain transparency, enabling traceability of digital design files and raw materials across global networks.

3.12.6 Sustainability and Energy Optimization

AI-driven energy management systems align AM production schedules with off-peak electricity rates, reducing power consumption by 22–27%. Fraunhofer Institute studies demonstrate that dynamic adjustments to laser intensity and cooling systems in metal AM lower energy use by 25%. Additionally, localized 3D printing eliminates carbon emissions associated with traditional logistics, as digital files can be transmitted globally for on-site production.

3.13 Conclusion

The implementation of AI and Industry 4.0 technologies in AM heralds a new era of intelligent, responsive production. By harnessing real-time data analytics, predictive algorithms, and interconnected systems, manufacturers achieve unprecedented levels of efficiency, customization, and sustainability. While challenges persist, the synergy between these innovations promises to redefine industrial paradigms, transforming AM from a prototyping tool into a cornerstone of global manufacturing. As industries continue to adopt these technologies, collaboration between policymakers, educators, and engineers will be essential to navigate ethical dilemmas and cultivate a workforce capable of steering this digital revolution.

AI's integration into AM processes delivers compound efficiency gains across design, production, and quality assurance domains. By automating iterative tasks, predicting system failures, and optimizing resource allocation, these technologies enable manufacturers to achieve resource efficiency, operational agility, cost saving, and quality assurance.

- **Resource Efficiency:** 30–45% reduction in material waste
- **Operational Agility:** 50–70% faster product development cycles.
- **Cost Savings:** 20–40% lower maintenance expenditures.
- **Quality Assurance:** >95% defect detection accuracy.

The synergistic combination of generative design algorithms, real-time monitoring systems, and predictive maintenance frameworks positions AI as an indispensable catalyst for AM's transition from a prototyping tool to full-scale production solution. As neural networks grow more sophisticated through multimodal learning techniques, future implementations promise near-autonomous AM ecosystems capable of self-optimization across global manufacturing networks.

References

Aeddula, O., Frank, M., Ruvald, R., Askling, C. J., Wall, J., & Larsson, T. (2024). AI-driven predictive maintenance for autonomous vehicles for product-service system development. *Procedia CIRP, 128*, 84–89. www.sciencedirect.com/science/article/pii/S2212827124006632

Ali, J., Kumar Singh, S., Jiang, W., Alenezi, A. M., Islam, M., Ibrahim Daradkeh, Y., *et al.* (2025). A deep dive into cybersecurity solutions for AI-driven IoT-enabled smart cities in advanced communication networks. *Computer Communications, 229*, 108000. www.sciencedirect.com/science/article/pii/S0140366424003475

Anantrasirichai, N., & Bull, D. (2022). Artificial intelligence in the creative industries: A review. *Artificial Intelligence Review, 55*(1), 589–656. https://doi.org/10.1007/s10462-021-10039-7

Bhaskaran, B., Nallasivam, J. D., & Aravinth, S. (2024). The art of material removal: An overview of subtractive manufacturing. *International Conference on Recent Advances in Science, Engineering, Technology and Management, Venue*; Sanjay Bhokare Group of Institutes, Miraj, India, MIRAJ, 22nd-23rd March, 2024.

Bousdekis, A., Papageorgiou, N., Magoutas, B., Apostolou, D., & Mentzas, G. (2017). A proactive event-driven decision model for joint equipment predictive maintenance and spare parts inventory optimization. *Procedia CIRP, 59*, 184–189. www.sciencedirect.com/science/article/pii/S2212827116309490

Brueckner, F., Mirko, R., Marquardt, F., Willner, R., Seidel, A., Thieme, S., *et al.* (2017, May 1). Process characteristics in high-precision laser metal deposition using wire and powder. *Journal of Laser Applications, 29*, 022301.

Chatziparaskeva, G., Papamichael, I., Voukkali, I., Loizia, P., Sourkouni, G., Argirusis, C., *et al.* (2022, July 20). End-of-life of composite materials in the framework of the circular economy. *Microplastics, 1*, 377–392.

da Silva, E. R., Shinohara, A. C., Nielsen, C. P., de Lima, E. P., & Angelis, J. (2020). Operating digital manufacturing in Industry 4.0: The role of advanced manufacturing technologies. *Procedia CIRP, 93*, 174–179. https://doi.org/10.1016/j.procir.2020.04.007

Da Silveira, G., Borenstein, D., & Fogliatto, F. S. (2001). Mass customization: Literature review and research directions. *International Journal of Production Economics, 72*(1), 1–13. www.sciencedirect.com/science/article/pii/S0925527300000797

Demartini, M., Tonelli, F., Pacella, M., & Papadia, G. (2021). A review of production planning models: Emerging features and limitations compared to practical implementation. *Procedia CIRP, 104*, 588–593. www.sciencedirect.com/science/article/pii/S2212827121009975

Dhatterwal, J. S., Kaswan, K. S., & Kumar, N. (2025). Self-powered integrated sensor system for printable fabrication. In R. K. Dhanaraj, P. Samuel, M. Sathyamoorthy, B. Balusamy, & V. B. Ravi (Eds.), *Advances in sensor systems* (pp. 187–235). Academic Press. www.sciencedirect.com/science/article/pii/B9780443137921000080

Di, L., & Yang, Y. (2022). Towards closed-loop material flow in additive manufacturing: Recyclability analysis of thermoplastic waste. *Journal of Cleaner Production, 362*, 132427. www.sciencedirect.com/science/article/pii/S0959652622020285

Dias, S., Espadinha-Cruz, P., & Matos, F. (2023). A Porter's Five Forces model proposal for additive manufacturing technology: A case study in Portuguese industry. *Procedia Computer Science, 217*, 165–176. https://doi.org/10.1016/j.procs.2022.12.906

Dou, C., Choudhary, H., Wang, Z., Baral, N. R., Mohan, M., Aguilar, R. A., *et al.* (2023). A hybrid chemical-biological approach can upcycle mixed plastic waste with reduced cost and carbon footprint. *One Earth, 6*(11), 1576–1590. www.sciencedirect.com/science/article/pii/S2590332223004906

Duflou, J. R., Sutherland, J. W., Dornfeld, D., Herrmann, C., Jeswiet, J., Kara, S., *et al.* (2012). Towards energy and resource efficient manufacturing: A processes and systems approach. *CIRP Annals, 61*(2), 587–609. www.sciencedirect.com/science/article/pii/S0007850612002016

Friesen, S., Bredfeldt, J., Kosak, T., *et al.* (2025). A prospectively deployed deep learning-enabled automated quality assurance tool for oncological palliative spine radiation therapy. *Lancet Digit Heal [Internet]*. 7(1):e13–22. Available from: www.sciencedirect.com/science/article/pii/S2589750024002437

Gan, M. X., & Wong, C. H. (2016). Practical support structures for selective laser melting. *Journal of Materials Processing Technology, 238*, 474–484. www.sciencedirect.com/science/article/pii/S0924013616302758

Gao, R. X., Wang, L., Helu, M., & Teti, R. (2020). Big data analytics for smart factories of the future. *CIRP Annals, 69*(2), 668–692. www.sciencedirect.com/science/article/pii/S0007850620301359

Gouiza, N., Jebari, H., & Reklaoui, K. (2024). Integration of IoT-enabled technologies and artificial intelligence in diverse domains: Recent advancements and future trends. *Journal of Theoretical and Applied Information Technology, 102*, 1975–2029.

Jiang, J., Xu, X., & Stringer, J. (2018, September 20). Support structures for additive manufacturing: A review. *Journal of Manufacturing and Materials Processing, 2*(4), 64.

Kanishka, K., & Acherjee, B. (2023). Revolutionizing manufacturing: A comprehensive overview of additive manufacturing processes, materials, developments, and challenges. *Journal of Manufacturing Processes, 107*, 574–619. www.sciencedirect.com/science/article/pii/S1526612523009726

Kehayias, C. E., Bontempi, D., Quirk, S., Friesen, S., Bredfeldt, J., Kosak, T., *et al.* (2025). A prospectively deployed deep learning-enabled automated quality assurance tool for oncological palliative spine radiation therapy. *The Lancet Digital Health, 7*(1), e13–e22. https://doi.org/10.1016/S2589-7500(24)00243-7

Korzenowski, A. L., Anzanello, M. J., Portugal, M. S., & ten Caten, C. (2013). Predictive models with endogenous variables for quality control in customized scenarios affected by multiple setups. *Computers & Industrial Engineering, 65*(4), 729–736. www.sciencedirect.com/science/article/pii/S036083521300140X

Kumar, P., Chauhan, S., & Awasthi, L. K. (2023). Artificial intelligence in healthcare: Review, ethics, trust challenges & future research directions. *Engineering Applications of Artificial Intelligence, 120*, 105894. www.sciencedirect.com/science/article/pii/S0952197623000787

Lettori, J., Raffaeli, R., Peruzzini, M., Schmidt, J., & Pellicciari, M. (2020). Additive manufacturing adoption in product design: An overview from literature and industry. *Procedia Manufacturing, 51*, 655–662. https://doi.org/10.1016/j.promfg.2020.10.090

Lu, Q. P., Wu, Y., Mao, X. D., Wan, H. J., Shao, J., Yu, Q. K., *et al.* (2024). Continuous quality improvement project to reduce the downtime of medical linear accelerators: A case study at Zhejiang Cancer Hospital. *Heliyon, 10*(9), e30668. www.sciencedirect.com/science/article/pii/S2405844024066994

Ma, J., Cao, B., Dong, S., Tian, Y., Wang, M., Xiong, J., *et al.* (2024). MLMD: A programming-free AI platform to predict and design materials. *npj Computational Materials, 10*(1), 59. https://doi.org/10.1038/s41524-024-01243-4

MajidiParast, S., Neamatian Monemi, R., & Gelareh, S. (2025). A graph convolutional network for optimal intelligent predictive maintenance of railway tracks. *Decision Analytics Journal, 14*, 100542. www.sciencedirect.com/science/article/pii/S2772662224001462

Martins, F. A. R., García-Bermejo, J. G., Casanova, E. Z., & Perán González, J. R. (2005). Automated 3D surface scanning based on CAD model. *Mechatronics, 15*(7), 837–857. www.sciencedirect.com/science/article/pii/S0957415805000371

Martins, R. F., Branco, R., Camacho, J., Macek, W., Marciniak, Z., Silva, A., *et al.* (2025). The influence of printing strategies on the fatigue crack growth behaviour of an additively manufactured Ti6Al4V Grade 23 titanium alloy. *International Journal of Fatigue, 197*, 108942. www.sciencedirect.com/science/article/pii/S0142112325001392

Mohapatra, A. (2024, November 8). Generative AI for predictive maintenance: Predicting equipment failures and optimizing maintenance schedules using AI. *International Journal of Scientific Research and Management, 12*, 1648–1672.

Morchid, A., Et-taibi, B., Oughannou, Z., Alami, R. El, Qjidaa, H., Jamil, M. O., *et al.* (2025). IoT-enabled smart agriculture for improving water management: A smart irrigation control using embedded systems and Server-Sent Events. *Scientific African, 27*, e02527. www.sciencedirect.com/science/article/pii/S2468227624004691

Munirathinam, S. (2020). Industry 4.0: Industrial Internet of Things (IIOT). In P. Raj & P. B. T. A. in C. Evangeline (Eds.), *The Digital Twin Paradigm for Smarter Systems and Environments: The Industry Use Cases* (pp. 129–164). Elsevier. www.sciencedirect.com/science/article/pii/S0065245819300634

Nguyen, H. D., Tran, K. P., Zeng, X., Koehl, L., & Castagliola, P. (2019). Industrial internet of things, big data, and artificial intelligence in the smart factory: A survey and perspective. *ISSAT International Conference on Data Science in Business, Finance and Industry*, Da Nang, Vietnam, July 2019.

Pagliari, C., Montalti, A., Frizziero, L., & Liverani, A. (2025). Enhancing ergonomic comfort: A study on customized cushion design using 3D scanning and additive manufacturing. *Results in Engineering, 25*, 104256. www.sciencedirect.com/science/article/pii/S259012302500341X

Pratap, B., Mohan, T. V. K., Amit, R. K., & Venugopal, S. (2024). Evaluating circular economy strategies for raw material recovery from end-of-life lithium-ion batteries: A system dynamics model. *Sustainable Production and Consumption, 50*, 191–204. www.sciencedirect.com/science/article/pii/S2352550924002197

Psarommatis, F. (2024). Zero defect manufacturing (ZDM): The future method for quality assurance in the era of Industry 4.0/5.0. In Martin A. Abraham (Ed.), *Encyclopedia of Sustainable Technologies* (Second Edition), Oxford: Elsevier (pp. 615–632). https://doi.org/10.1016/B978-0-323-90386-8.00127-3

Puttemans, I., Caris, A., & Braekers, K. (2025). The integration of location and inventory decisions in supply chain networks: A literature review and future prospects. *Transportation Research Part E: Logistics and Transportation Review, 195*, 103970. www.sciencedirect.com/science/article/pii/S1366554525000110

Qu, P. (2024). Design of supply chain logistics intelligent management information system based on GIS optimization model. *Procedia Computer Science, 243*, 396–405. https://doi.org/10.1016/j.procs.2024.02.056

Rajak, P., Ganguly, A., Adhikary, S., & Bhattacharya, S. (2023). Internet of Things and smart sensors in agriculture: Scopes and challenges. *Journal of Agriculture and Food Research, 14*, 100776. www.sciencedirect.com/science/article/pii/S2666154323002831

Rashid, A. Bin, & Kausik, M. D. A. K. (2024). AI revolutionizing industries worldwide: A comprehensive overview of its diverse applications. *Hybrid Advances, 7*, 100277. www.sciencedirect.com/science/article/pii/S2773207X24001386

Saha, D., Padhiary, M., & Chandrakar, N. (2025). AI vision and machine learning for enhanced automation in food industry: A systematic review. *Food and Humanities, 4*, 100587. https://doi.org/10.1016/j.foh.2025.100587

Sahoo, S., & Lo, C. Y. (2022). Smart manufacturing powered by recent technological advancements: A review. *Journal of Manufacturing Systems, 64*, 236–250. www.sciencedirect.com/science/article/pii/S0278612522001042

Sani, A. R., Zolfagharian, A., & Kouzani, A. Z. (2024). Artificial intelligence-augmented additive manufacturing: Insights on closed-loop 3D printing. *Advanced Intelligent Systems, 6*(10), 2400102. https://doi.org/10.1002/aisy.202400102

Shiferaw, M. Z., & Gebremedhen, H. S. (2022). Recycled polymer for FDM 3D printing filament material: Circular economy for sustainability of additive manufacturing. In M. L. Berihun (Ed.), *Advances of Science and Technology* (pp. 243–261). Cham: Springer International Publishing.

Shukla, A., Ross, R., Bhattacharya, B., & Stumpf, A. (2025). Autonomous water sampling and quality monitoring in remote locations: A novel approach using a remote-controlled boat. *HardwareX, 22*, e00634. www.sciencedirect.com/science/article/pii/S2468067225000124

Soares, C. T. de M., Ek, M., Östmark, E., Gällstedt, M., & Karlsson, S. (2022). Recycling of multi-material multilayer plastic packaging: Current trends and future scenarios. *Resources, Conservation and Recycling, 176*, 105905. www.sciencedirect.com/science/article/pii/S0921344921005140

Soori, M., Arezoo, B., & Dastres, R. (2023). Internet of Things for smart factories in Industry 4.0: A review. *Internet of Things and Cyber-Physical Systems, 3*, 192–204. www.sciencedirect.com/science/article/pii/S2667345223000275

Soori, M., Jough, F. K. G., Dastres, R., & Arezoo, B. (2025). Additive manufacturing modification by artificial intelligence, machine learning, and deep learning: A review. *Additive Manufacturing Frontiers*, 200198. www.sciencedirect.com/science/article/pii/S2950431725000085

Svetlizky, D., Das, M., Zheng, B., Vyatskikh, A. L., Bose, S., Bandyopadhyay, A., *et al.* (2021). Directed energy deposition (DED) additive manufacturing: Physical characteristics, defects, challenges and applications. *Materials Today, 49*, 271–295. www.sciencedirect.com/science/article/pii/S1369702121001139

Taher, M. A., & Bashar, M. (2024, June 7). The impact of lean manufacturing concepts on industrial processes' efficiency and waste reduction. *International Journal of Progressive Research in Engineering and Management Sciences, 4*, 338–349.

Tuli, N. T., Khatun, S., & Rashid, A. Bin. (2024, March 15). Unlocking the future of precision manufacturing: A comprehensive exploration of 3D printing with fiber-reinforced composites in aerospace, automotive, medical, and consumer industries. *Heliyon, 10*(5), e27328. https://doi.org/10.1016/j.heliyon.2024.e27328

Ullah, H., Uzair, M., Jan, Z., & Ullah, M. (2024). Integrating Industry 4.0 technologies in defense manufacturing: Challenges, solutions, and potential opportunities. *Array, 23*, 100358. https://doi.org/10.1016/j.array.2024.100358

van den Berg, M., Neumann, K., van Vuuren, D. P., Bouwman, A. F., Kram, T., & Bakkes, J. (2016). Exploring resource efficiency for energy, land and phosphorus use: Implications for resource scarcity and the global environment. *Global Environmental Change, 36*, 21–34. www.sciencedirect.com/science/article/pii/S0959378015300534

Veza, I., Asy'ari, M. Z., Idris, M., Epin, V., Rizwanul Fattah, I. M., & Spraggon, M. (2023). Electric vehicle (EV) and driving towards sustainability: Comparison between EV, HEV, PHEV, and ICE vehicles to achieve net zero emissions by 2050 from EV. *Alexandria Engineering Journal, 82*, 459–467. www.sciencedirect.com/science/article/pii/S1110016823009055

Vlah, D., Žavbi, R., & Vukašinović, N. (2020, May 1). Evaluation of topology optimization and generative design tools as support for conceptual design. *Proceedings of the Design Society: Design Conference, 1*, 451–460.

Walmsley, T. G., Ong, B. H. Y., Klemeš, J. J., Tan, R. R., & Varbanov, P. S. (2019). Circular integration of processes, industries, and economies. *Renewable and Sustainable Energy Reviews, 107*, 507–515. www.sciencedirect.com/science/article/pii/S1364032119301753

Wang, Z., Li, J. S., Pan, H. R., Wu, J. Y., & Yan, W. A. (2025). Research on multimodal generative design of product appearance based on emotional and functional constraints. *Advanced Engineering Informatics, 65*, 103106. https://doi.org/10.1016/j.aei.2024.103106

Yi, F., Zhou, Q., Wang, C., Yan, Z., & Liu, B. (2021). Effect of powder reuse on powder characteristics and properties of Inconel 718 parts produced by selective laser melting. *Journal of Materials Research and Technology, 13*, 524–533. www.sciencedirect.com/science/article/pii/S223878542100435X

Zheng, C., Li, Z., Janardhanan, M., Zhang, Z., & Zhang, L. (2025). Balancing and scheduling human-robot collaborative assembly lines with heterogeneous robots and limited resources: Constraint programming approach and fruit fly optimization algorithm. *Computers & Industrial Engineering, 203*, 111046. https://doi.org/10.1016/j.cie.2025.111046

4

Investigation on Damper Material While Designing and Simulation of Low-Cost Magnetorheological Fluid: A Manufacturing Systems-Based Case Study

Md Umar Ibrahim, Aseeb Sabu, K. P. Lijesh, and S. M. Muzakkir

4.1 Introduction

The magnetorheological (MR) dampers are crucial in mitigating vibrations and minimizing bounce in vehicles. Persistent exposure to these vibrations and bumps can affect the human body, potentially leading to work-related musculoskeletal disorders [1]. Despite technological advancements, conventional suspension systems – known as passive suspension systems – are still used in budget-friendly cars. Passive suspension systems are valued for their simplicity, cost-effectiveness, and reliability. They offer consistent handling and lower costs but lack real-time adjustability and automatic adaptation to road conditions. To address these limitations, active and semi-active suspension systems [2] are used but are cost prohibitive in low-cost vehicles. The present research work is aimed to design and simulate a low-cost semi-active MR suspension system by modifying materials and fluids for the MR damper. This approach seeks to enhance comfort and performance without the excessive costs associated with fully active systems. Rashid et al. [3] delve into the practical application of MR dampers for car suspension control, emphasizing the utilization of MR fluids to achieve controllable damping. In 2002, Delphi Automotive Corporation introduced MagneRide™ suspensions featuring MR dampers, now used in vehicles by Audi, Ferrari, General Motors, and Holden Special Vehicles (LORD Corporation product information). This makes suitable for high-performance applications [4]–[7]. Previous research has demonstrated the potential of MR brakes in various applications [8]–[13], but a magnetorheological brake (MRB) that provides a better performance needs to be explored. And there are challenges in optimizing their design for maximum efficiency and reliability due to issues in the MR fluid stability,

DOI: 10.1201/9781003506492-4

particle size distribution, gap between disk and brake pad, misalignment of shaft, precise magnetic field control, and heat dissipation [14]–[23]. The results concluded that MR dampers significantly outperformed passive suspensions in terms of ride comfort, handling, and road-holding stability but because of high cost it is not feasible to implement it on lower segment vehicles.

4.2 Methodology

The methodology adopted is a multifaceted strategy aimed at optimizing every aspect of the damper's design, manufacturing, and performance as detailed below.

I. **Analysis of LORD MR Damper**

   Examine the components and pricing structure of the LORD MR damper to identify key parameters that influence both performance and cost.

II. **Design and Simulation of MR Damper**

   Using the parameters from the LORD MR damper as a reference baseline, simulation of MR damper is carried out to gain insights into performance and cost reduction strategies.

III. **Market Research**

   Conduct comprehensive market research to identify a cost-effective MR fluid. Explore potential materials for the damper cylinder and piston, considering factors such as cost, availability, and machining requirements.

IV. **Material Selection**

   Identify appropriate materials for the MR damper cylinder and piston, and compare the costs associated with each.

V. **Cost Calculation and Comparison**

   Calculate the overall cost of the MR damper utilizing the newly identified MR fluid and each of the three selected materials. Analyse the cost variations to determine the most economical option.

4.2.1 Analysis of MR Damper

The LORD RD-8040-1 [24] dampers are small, MR fluid-based dampers designed for industrial suspension systems (Figure 4.1). These dampers provide adjustable damping levels by enhancing the MR fluid's yield strength based on the magnetic field intensity. The cost of these dampers is

FIGURE 4.1
LORD MR Damper [24].

approximately ₹42,000 [24]. This pricing reflects the advanced technology and performance capabilities integrated into these dampers, making them suitable for demanding industrial applications where precise damping control is essential. Table 4.1 shows the dimension of MR damper.

These dimensions are crucial for ensuring compatibility, performance, and integration of MR dampers in industrial and automotive systems [25] as shown in Figures 4.2 and 4.3.

The LORD damper utilizes the MRF 122 magnetorheological fluid (MRF) with following properties as shown in Table 4.2.

The LORD RD-8040-1 damper typically features specific materials for its cylinder and piston components to ensure durability, performance, and compatibility with magneto-rheological (MR) fluid damping systems. While exact material specifications can vary based on specific models and design

TABLE 4.1

Dimension of MR Damper [24]

SL.No.	Name	Expression (mm)	Description
1	R	6	Piston rod radius
2	H	10	Radial distance from piston rod to
3	T	7	coil Width thickness of cylinder
4	R	23	Piston radius
5	L	22	Distance between poles
6	L	28	Pole length
7	H	0.7	Fluid gap
8	D	7	Coil width

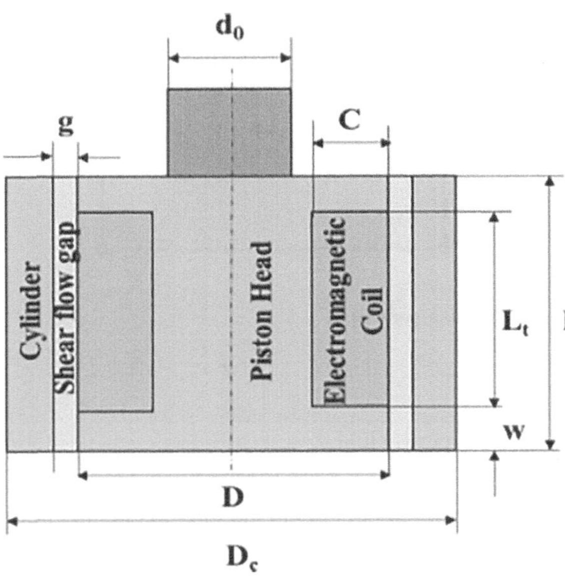

FIGURE 4.2
Schematic diagram of MR damper [25].

considerations. The cylinder body of the damper is commonly made from aluminium or aluminium alloy. The piston within the damper is often made from hardened steel or stainless steel. The electromagnetic coil in the LORD RD-8040-1 damper plays a crucial role in generating the magnetic field that controls the behaviour of the MR fluid. The core of the electromagnetic coil may be made from ferromagnetic materials such as iron or steel. These materials enhance the magnetic flux density within the coil when an electrical current passes through, resulting in a stronger magnetic field.

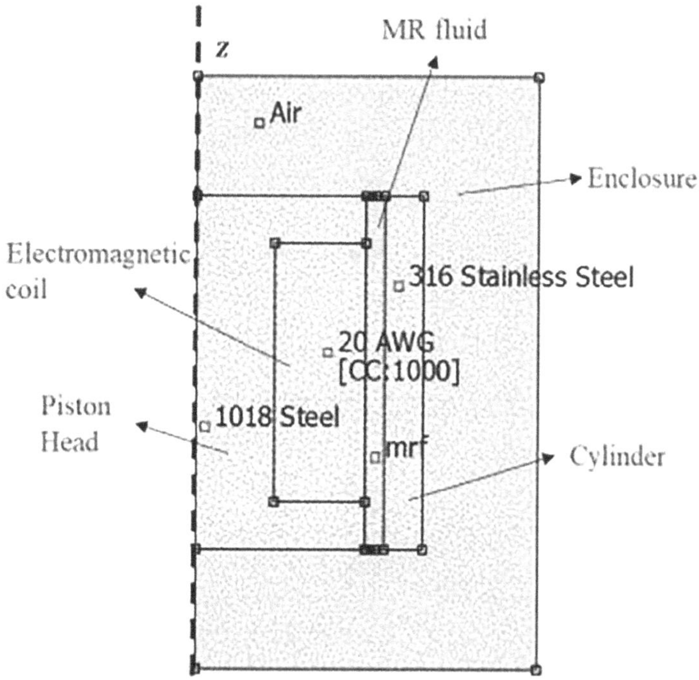

FIGURE 4.3
Asymmetry diagram of MR damper [25].

TABLE 4.2
MR Fluid Properties [24]

Parameter	Value
1. Viscosity (Pa-s) at 40 C	0.092 ±+- 0.015
2. Density (g/cm3)	2.98 – 3.18
3. Solid Content by wt., %	72%
4. Weight(g)	800

4.2.1.1 Cost Analysis of LORD MR Damper

The costs for the LORD MR damper, particularly the MRF 122 fluid, as well as the expenses for the other three critical components (cylinder, piston, and coil), and the associated manufacturing costs, are significant. Additionally, factors such as material quality, precision engineering, and advanced manufacturing techniques contribute to the overall cost (see Table 4.3).

Based on the analysis conducted, it is evident that a significant portion of the cost associated with manufacturing MR dampers lies in the procurement of the MRF 122 fluid.

TABLE 4.3
Costing of MR Damper Components [24]

Component	Cost (Rs)
1. MRF 122 Fluid	29,000
2. Cylinder, piston, coil, and manufacturing	13,000
Total	42,000

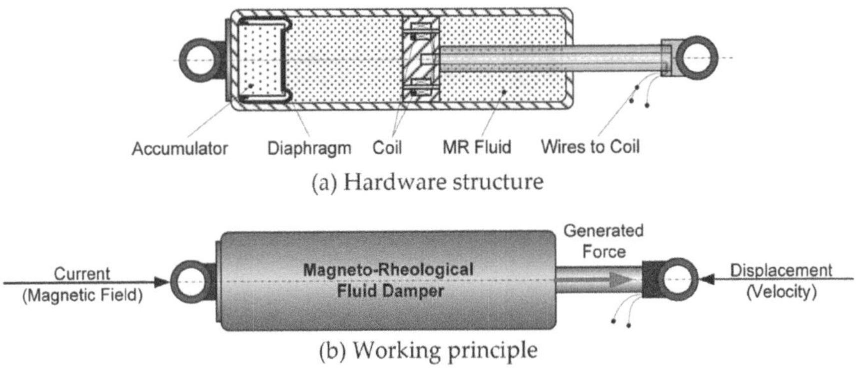

FIGURE 4.4
Working of MR damper [26].

4.2.2 Design, Analysis, and Simulation

MR dampers are a type of shock absorber that use a special fluid known as MR fluid to control vibrations. In an MR damper, the operating principle varies depending on the presence or absence of a magnetic field. Without a magnetic field, the MR fluid behaves like a regular viscous fluid. As the piston moves within the damper's cylinder, the MR fluid flows through the gap between the piston and cylinder, allowing for controlled movement (see Figure 4.4). The alignment forms chains of particles that restrict the flow of the MR fluid, effectively increasing its viscosity (see Figures 4.5, 4.6 and 4.7).

A 3D model of the MR damper, as per dimensions and material specifications, is created (see Figure 4.8).

In this study, an axis-symmetric model in ANSYS is utilized. The geometry, specifications, and materials used are identical to those of the RD-1005-3 MR damper manufactured by LORD Corporation.

Figure 4.9 illustrates the magnetic field intensity and the electromagnetic field distribution around the coil of the LORD MR damper.

4.2.3 Market Research

Through comprehensive market research, an MR fluid sourced from ARUS MR TECH was identified. This MR fluid demonstrates a marginally reduced

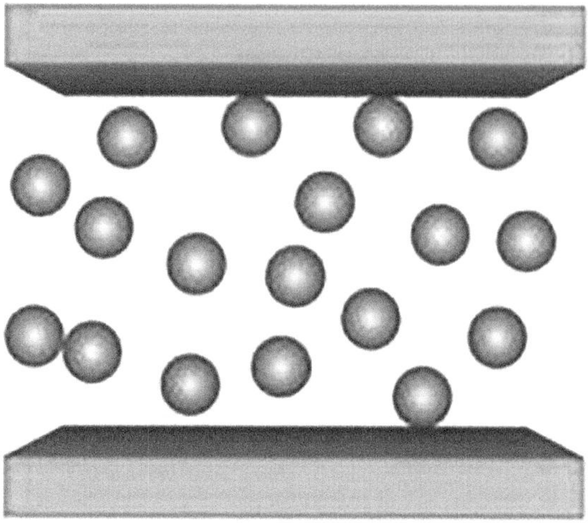

FIGURE 4.5
Without magnetic field [26].

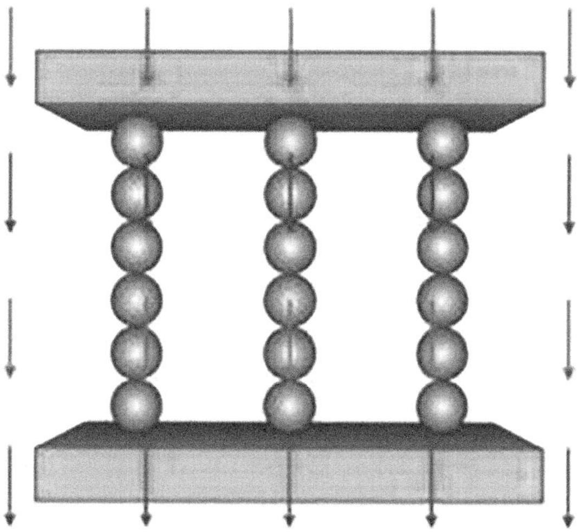

FIGURE 4.6
With magnetic field [26].

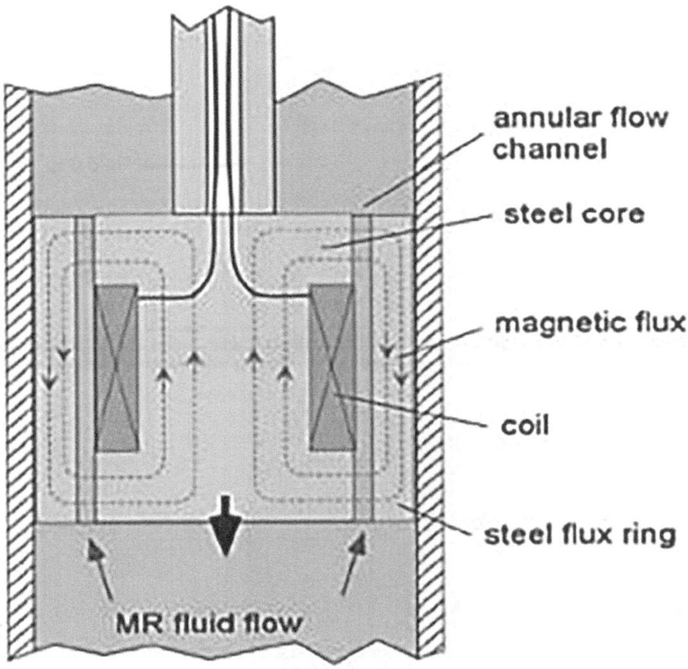

FIGURE 4.7
Magnetic field lines [26].

cost of ₹9,900 compared to the LORD MR fluid while retaining exceptional performance attributes.

The commercial availability of this MR fluid ensures compatibility with MR damper production requirements. Its comparable rheological properties and magnetic responsiveness make it a viable alternative, allowing to achieve substantial reductions in manufacturing costs without compromising the damping efficiency or overall quality of MR dampers. Figure 4.10 shows the ARUS MR Fluid.

Arus MR fluid is a type of MR fluid that exhibits unique technical properties, particularly in the context of its application in damping systems like MR dampers. Some of the key technical properties of Arus MR fluid are given in Table 4.4 and Figures 4.11, 4.12, and 4.13.

4.2.4 Material Selection

The material for piston and cylinder must exhibit appropriate magnetic permeability to effectively interact with the MR fluid under varying magnetic fields, ensuring optimal damping performance. The mechanical strength is essential to withstand operational loads and pressures without deformation

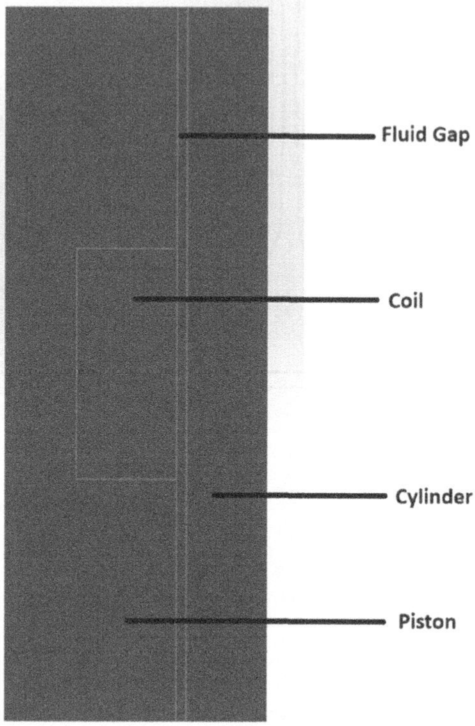

FIGURE 4.8
MR damper CAD model.

or failure, guaranteeing the damper's structural integrity. Additionally, corrosion resistance is crucial to prolong the components' lifespan, especially when exposed to harsh environmental conditions that may accelerate deterioration. Table 4.5 shows the raw material cost.

The machining operations were optimized for dimensional accuracy and tolerances while considering per-hour machining costs for budgeting and cost estimation. Table 4.6 details the manufacturing process and machining of the cylinder for an MR damper, including the associated cost per hour according to the MSME tool room.

Table 4.7 outlines the manufacturing procedures for the piston along with the corresponding hourly costs according to the MSME tool room.

Following the manufacturing processes for the cylinder and piston of an MR damper, the subsequent steps involve assembly and testing. The assembly process entails fitting the machined cylinder and piston together according to design specifications, ensuring precision and alignment. Subsequently, the assembled MR damper undergoes rigorous testing procedures to validate its

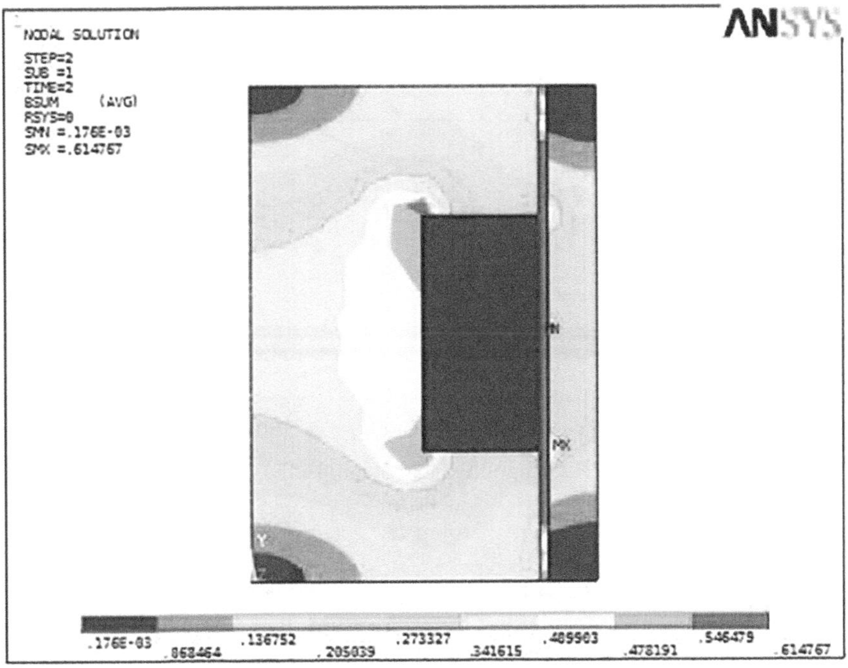

FIGURE 4.9
Magnetic flux density Ansys model.

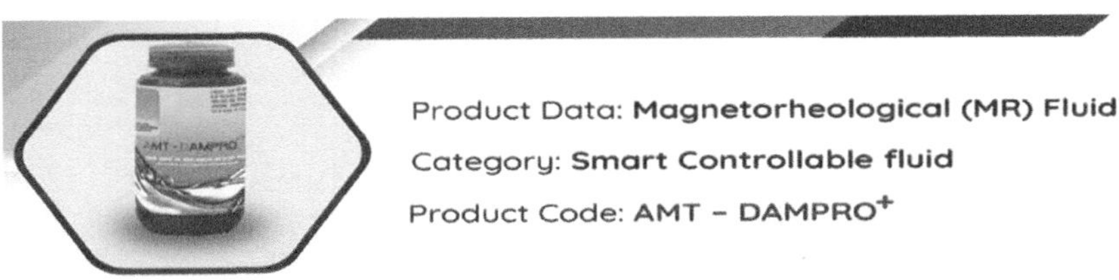

FIGURE 4.10
ARUS MR fluid.

TABLE 4.4
Properties of Arus MR Fluid

Appearance	Dark Gray Liquid
Density (g/cm^3)	2.45
Viscosity (Pa.s) 40°c	0.29
Solid content by weight %	73

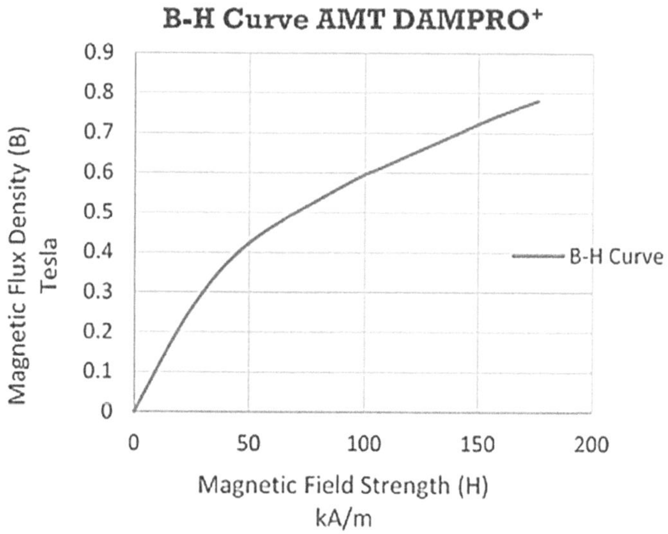

FIGURE 4.11
B-H curve of ARUS MR fluid.

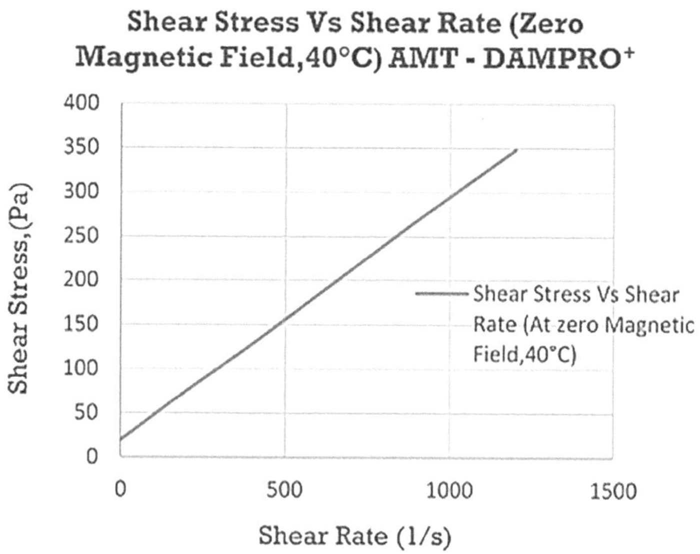

FIGURE 4.12
Shear stress vs shear rate of ARUS MR fluid.

functionality, damping performance, and overall quality. These tests include dynamic performance evaluations, leakage tests, and endurance testing to ensure optimal performance and reliability under varying operating conditions. Table 4.8 presents the assembly process and testing.

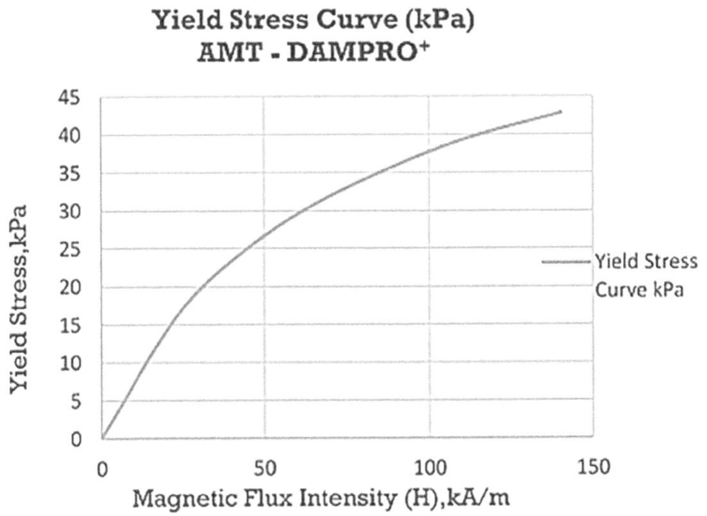

FIGURE 4.13
Yield stress curve of ARUS MR fluid.

TABLE 4.5
Raw Material Cost

Material	Hollow (Rs/kg)	Solid (Rs/kg)
316 stainless steel	124	295
AISI 4140	500	100
Aluminium 6063	240	295

4.2.4.1 Comparison of Materials

In this section, different materials and costs for the MR damper are compared. A critical consideration is the selection of materials. Specifically, the materials must be able to withstand the induced forces and pressure, resist corrosion, provide the desired magnetic properties, and possess a low mass density. Aluminium 6063 is a popular choice for making the cylinder and piston of an MR damper due to several advantageous properties that make it particularly well-suited for this application. Its combination of mechanical strength, corrosion resistance, and favourable machinability ensures that the components manufactured from this alloy are durable and perform reliably under operational stresses. Additionally, Aluminium 6063 provides the desired magnetic properties by being non-magnetic, which is crucial for the effective functioning of the MR damper. This material's lightweight nature contributes to the overall reduction in the weight of the damper, which is essential for applications in industries such as automotive

TABLE 4.6

Process of Manufacturing of Cylinder

Process	Machine	Time (hour)	Price (Rs)
1. Material Preparation			
Cutting Machines	Band Saws: For cutting the hollow cylinder material to the required length.	1–2	80–160
2. Machining			
Turning	CNC Lathe: For turning the external surface of the hollow cylinder to achieve the required dimensions and finish.	2–3	400–600
Boring	Horizontal Boring Machine: For enlarging and finishing the internal diameter of the hollow cylinder.	1–2	450–900
Milling	CNC Milling Machine: For creating necessary slots or additional features on the cylinder.	1–2	400–800
3. Heat Treatment			
Heat Treatment Furnaces	Vacuum Furnace: For hardening and tempering the cylinder to improve its mechanical properties.	4–6	200
4. Surface Finishing			
Grinding	Cylindrical Grinder: For precision grinding of both the external and internal surfaces to achieve a smooth finish.	2–3	900–1,250
Polishing	Polishing Machine: For achieving a high-quality surface finish to reduce friction.	1	500
Honing	Horizontal or Vertical Honing Machines: These machines are ideal for achieving the final finish on the internal surfaces of the cylinder. Honing can produce extremely smooth finishes and precise bore sizes, often achieving Ra values below 0.2 μm.	2–3	800
Lapping	Lapping Machines: These are used to further refine the surface finish after grinding and honing. Lapping can produce an ultra-smooth surface, reducing Ra values to 0.1 μm or even lower if needed	1–2	300–600
5. Coating			
Plating Equipment	Electroplating System: For applying a protective coating to the internal and external surfaces.	2–3	500

and aerospace where weight savings are paramount. These characteristics collectively make Aluminium 6063 a highly favourable material for manufacturing the critical components of MR dampers [7]. Table 4.9 shows technical Properties of Aluminium 6063.

TABLE 4.7
Process of Manufacturing Piston

Process	Machine	Time (Hour)	Price (Rs)
1. Material Preparation			
Cutting Machines	Band Saws: For cutting the hollow cylinder material to the required length.	1–2	80–160
	Laser Cutters: For precise cutting and initial shaping	1–2	80–160
2. Machining			
Turning	CNC Lathe: To turn the external surface of the solid piston to achieve precise dimensions and surface finish	2–3	400–600
Milling	CNC Milling Machine: For creating any necessary slots or flat surfaces on the piston, such as mounting features or seal grooves.	1–2	400–800
3. Heat Treatment			
Heat Treatment Furnaces	Vacuum Furnace: For hardening and tempering the piston to enhance its strength and durability.	4–6	200
4. Surface Finishing			
Grinding	Surface Grinder: For achieving precise dimensions and a smooth surface finish on the piston.	2–3	250–375
Superfinishing	Superfinishing Machines: These machines are particularly effective for achieving a very fine surface finish on the piston. Superfinishing can achieve Ra values as low as 0.05 µm, ensuring an exceptionally smooth surface	1–2	1,000–2,000
Polishing	Polishing Machine: For achieving a high-quality surface finish to reduce friction.	1	500
Electropolishing	Electropolishing: For metal pistons, this electrochemical process can remove microscopic peaks and valleys from the surface, resulting in a highly smooth and reflective finish	1–2	500
5. Coating			
Plating Equipment	Electroplating System: For applying a protective coating to the internal and external surfaces.	2–3	500

4.2.4.2 Cost Estimation of MR Damper

Calculating the cost of an MR damper using Aluminium 6063 for the cylinder and piston involves several steps, and the process includes determining the price of the raw Aluminium 6063, machining costs for both the cylinder and piston, and the assembly expenses. By accounting for each of these factors,

TABLE 4.8

Assembly Process and Testing

Process	Machine	Time	Price (Rs)
1. **Welding and Bonding**	Ultrasonic Welding Machine: For joining metal parts	1–2	400
2. **Testing and Calibration**	Dynamic Testing Machine (Dynamometer): Simulates operating conditions to test performance and functionality	1–2	1,000
3. **Quality Control and Inspection**	CMM (Coordinate Measuring Machine): For precise dimensional inspection	2–4	900–1,250

TABLE 4.9

Technical Properties of Aluminium 6063

Property	Value
Density	2.70 g/cm^3
Tensile Strength	145 MPa (21,000 psi)
Yield Strength	90 MPa (13,000 psi)
Modulus of Elasticity	68.9 GPa (10,000 ksi)

TABLE 4.10

Aluminium 6063 MR Damper Cost

Process	Cost approx. (Rs)
Raw Material	544
Manufacturing	9,000
Assembling and Testing	2,400
MR fluid	9,900
Total	21,800

the total cost of producing the MR damper can be approximately calculated (Table 4.10).

The 316 stainless steel is an excellent choice for manufacturing MR damper cylinders and pistons due to its exceptional properties. It offers superior corrosion resistance, high mechanical strength, and sufficient magnetic properties for MR applications (Table 4.11). Additionally, it is easy to machine, allowing for precise manufacturing of complex shapes, and maintain its mechanical properties across a wide range of temperatures. While not the lightest material, its balance of strength, durability, and cost-effectiveness makes it a reliable and effective option for MR damper components.

Calculating the cost of an MR damper using 316 stainless steel for both the cylinder and piston involves several steps. First, the cost of the raw material is determined by sourcing 316 stainless steel. Next, the manufacturing

TABLE 4.11

Technical Properties 316 Stainless Steel

Property	Value
Density	8.00 g/cm^3
Tensile Strength	515 MPa (21,000 psi)
Yield Strength	205 MPa (13,000 psi)
Modulus of Elasticity	193 GPa (10,000 ksi)
Magnetic Permeability	1.02

TABLE 4.12

316 Stainless Steel MR Damper Cost

Process	Cost Approx. (Rs)
Raw material	1,383
Manufacturing	7,000
Assembling and testing	2,400
MR fluid	9,900
Total	20,683

TABLE 4.13

Technical Properties AISI 4140

Property	Value
Density	7.85 g/cm^3
Tensile Strength	655 MPa (21,000 psi)
Yield Strength	415 MPa (13,000 psi)
Modulus of Elasticity	205 GPa (10,000 ksi)

costs are calculated, including cutting, shaping, and finishing the 316 stainless steel to the required dimensions for the cylinder and piston. The high machinability of 316 stainless steel allows for efficient manufacturing processes, further minimizing costs. By considering the raw material cost, reduced machining requirements, and efficient assembly, the total cost of producing the MR damper using 316 stainless steel can be approximately estimated. This approach ensures a high-quality, cost-effective end product (Table 4.12).

AISI 4140 is an ideal material for manufacturing MR damper cylinders and pistons due to its outstanding mechanical properties and wear resistance. AISI 4140 is known for its good machinability, allowing for precise manufacturing, and achieving necessary tolerances and surface finishes. While not as corrosion resistant as stainless steel, its performance can be improved with protective coatings or surface treatments (Table 4.13). The cost-effectiveness of AISI 4140, coupled with its availability in various forms,

TABLE 4.14

AISI 4140 MR Damper Cost

Process	Cost Approx. (Rs)
Raw material	1,323
Manufacturing	8,000
Assembling and testing	2400
MR fluid	9,900
Total	21,600

makes it a practical and economical choice for producing high-quality MR damper components, ensuring reliable and efficient performance in various applications.

Calculating the cost of an MR damper using AISI 4140 for the cylinder and piston involves several steps. First, determine the raw material cost of AISI 4140 alloy steel. Next, calculate the manufacturing costs, including machining, heat treatment, and finishing processes. Notably, due to its properties, AISI 4140 does not require extensive machining for high surface roughness, reducing manufacturing time and costs. Finally, combine these costs to obtain the total cost of the MR damper (Table 4.14).

4.3 Results and Discussion

In this research work, the primary goal was to develop a cost-effective MR damper by analysing existing high-performance models and identifying opportunities for cost reduction through material selection and MR fluid optimization. Initially, the LORD MR damper was thoroughly analysed to understand its components, functionality, and pricing. It was identified that the choice of materials for the cylinder and piston, along with the selection of MR fluid, significantly impacts the overall cost of the damper. After conducting market research, a suitable MR fluid was identified at a cost of ₹9,900 per litre. This fluid meets the necessary performance criteria while offering a more affordable alternative to the existing options. Concurrently, three materials were evaluated for the cylinder and piston: Aluminium 6063, 316 Stainless Steel, and AISI 4140. The cost analysis of manufacturing the MR damper with Aluminium 6063, 316 Stainless Steel, and AISI 4140 yielded ₹21,800, ₹20,600, and ₹21,600, respectively. It is evident that 316 Stainless Steel is the optimal choice for achieving a low-cost MR damper.

4.4 Conclusion

In this research work, an in-depth analysis of the LORD MR damper was conducted to understand its components and pricing structure. Utilizing this knowledge, simulation of the MR damper with the same parameters as the LORD MR damper was carried out to identify opportunities for cost reduction. Market research provided an MR fluid priced at ₹9,900 per litre, and three materials suitable for the MR damper cylinder were identified: Aluminium 6063, 316 Stainless Steel, and AISI 4140. After calculating the total cost of the MR damper with the new fluid and materials, it was observed that using Aluminium 6063 resulted in a cost of ₹21,800, 316 Stainless Steel resulted in a cost of ₹20,600, and AISI 4140 resulted in a cost of ₹21,600. Among these, 316 Stainless Steel emerged as the most cost-effective option, primarily due to its lower machining requirements.

References

[1] Y. L. Chen, H. Alexander, and Y. M. Hu, "Self-reported musculoskeletal disorder symptoms among bus drivers in the Taipei metropolitan area," *Int. J. Environ. Res. Public Health*, vol. 19, no. 17, 2022, doi: 10.3390/ijerph191710596.

[2] D. Ivers and D. LeRoy, "Improving vehicle performance and operator ergonomics: Commercial application of smart materials and systems," *J. Intell. Mater. Syst. Struct.*, vol. 24, no. 8, pp. 903–907, 2013, doi: 10.1177/1045389X12445630.

[3] M. M. Rashid, M. M. Ferdaus, M. H. Hasan, and A. Rahman, "ANSYS finite element design of an energy saving magneto-rheological damper with improved dispersion stability," *J. Mech. Sci. Technol.*, vol. 29, no. 7, pp. 2793–2802, 2015, doi: 10.1007/s12206-015-0608-x.

[4] D. J. Carlson and M. R. Jolly, "MR fluid, foam and elastomer devices," *Mechatronics*, vol. 10, pp. 555–569, 2000, doi: 10.1016/0375-9601(86)90214-8.

[5] S. G. Sherman and N. M. Wereley, "Effect of particle size distribution on chain structures in magnetorheological fluids," *IEEE Trans. Magn.*, vol. 49, no. 7, pp. 3430–3433, 2013, doi: 10.1109/TMAG.2013.2245409.

[6] C. Sarkar and H. Hirani, "Effect of particle size on shear stress of magnetorheological fluids," *Smart Sci.*, vol. 3, no. 2, pp. 65–73, 2015, doi: 10.1080/23080477.2015.11665638.

[7] S. S. Goilkar and H. Hirani, "Parametric study on balance ratio of mechanical face seal in steam environment," *Tribol. Int.*, vol. 43, no. 5–6, pp. 1180–1185, 2010, doi: 10.1016/j.triboint.2009.12.019.

[8] K. P. Lijesh, D. Kumar, and H. Hirani, "Effect of disc hardness on MR brake performance," *Eng. Fail. Anal.*, vol. 74, pp. 228–238, 2017, doi: 10.1016/j.engfailanal.2017.01.009.

[9] V. K. Sukhwani and H. Hirani, "A comparative study of magnetorheological-fluid-brake and magnetorheological-grease-brake," *Tribol. Online*, vol. 3, no. 1, pp. 31–35, 2008, doi: 10.2474/trol.3.31.

[10] H. Hirani and C. S. Manjunatha, "Performance evaluation of a magnetorheological fluid variable valve," *Proc. Inst. Mech. Eng. Part D J. Automob. Eng.*, vol. 221, no. 1, pp. 83–93, 2007, doi: 10.1243/09544070JAUTO408.
[11] V. K. Sukhwani, L. Vijaya, and H. Hirani, "Performance evaluation of MR brake: An experimental study," *Indian J. Tribol.*, pp. 67–52, 2006.
[12] S. M. Muzakkir, H. Hirani, and G. D. Thakre, "Lubricant for heavily loaded slow-speed journal bearing," *Tribol. Trans.*, vol. 56, no. 6, pp. 1060–1068, Nov. 2013, doi: 10.1080/10402004.2013.823530.
[13] S. M. Muzakkir, K. P. Lijesh, H. Hirani, and G. D. Thakre, "Effect of cylindricity on the tribological performance of heavily loaded slow-speed journal bearing," *Proc. Inst. Mech. Eng. Part J J. Eng. Tribol.*, vol. 229, no. 2, 2015, doi: 10.1177/1350650114548053.
[14] H. Hirani, "Multiobjective optimization of journal bearing using mass conserving and genetic algorithms," *Proc. Inst. Mech. Eng. Part J J. Eng. Tribol.*, vol. 219, no. 3, pp. 235–248, 2005, doi: 10.1243/135065005X9844.
[15] H. Hirani, K. Athre, and S. Biswas, "Lubricant shear thinning analysis of engine journal bearings," *Tribol. Trans.*, vol. 44, no. 1, pp. 125–131, 2001, doi: 10.1080/10402000108982435.
[16] H. Hirani, K. Athre, and S. Biswas, "Rapid and globally convergent method for dynamically loaded journal bearing design," *Proc. Inst. Mech. Eng. Part J J. Eng. Tribol.*, vol. 212, no. 3, pp. 207–213, 1998, doi: 10.1243/1350650981542010.
[17] H. Hirani, K. Athre, and S. Biswas, "Comprehensive design methodology for an engine journal bearing," *Proc. Inst. Mech. Eng. Part J J. Eng. Tribol.*, vol. 214, no. 4, pp. 401–412, 2000, doi: 10.1243/1350650001543287.
[18] P. Kumar, H. Hirani, and A. Kumar Agrawal, "Effect of gear misalignment on contact area: Theoretical and experimental studies," *Measurement*, vol. 132, pp. 359–368, Jan. 2019, doi: 10.1016/J.MEASUREMENT.2018.09.070.
[19] K. P. Lijesh, S. M. Muzakkir, and H. Hirani, "Failure mode and effect analysis of passive magnetic bearing," *Eng. Fail. Anal.*, vol. 62, pp. 1–20, Apr. 2016, doi: 10.1016/j.engfailanal.2015.11.033.
[20] H. Hirani, "Root cause failure analysis of outer ring fracture of four-row cylindrical roller bearing," *Tribol. Trans.*, vol. 52, no. 2, pp. 180–190, 2009, doi: 10.1080/10402000802180151.
[21] K. P. Lijesh and H. Hirani, "Magnetic bearing using rotation magnetized direction configuration," *J. Tribol.*, vol. 137, no. 4, 2015, doi: 10.1115/1.4030344.
[22] K. P. Lijesh and H. Hirani, "Design and development of Halbach electromagnet for active magnetic bearing," *Prog. Electromagn. Res. C*, vol. 56, pp. 173–181, 2015.
[23] S. S. Goilkar and H. Hirani, "Design and development of a test setup for online wear monitoring of mechanical face seals using a torque sensor," *Tribol. Trans.*, vol. 52, no. 1, pp. 47–58, 2009, doi: 10.1080/10402000802163017.
[24] Lord. "Lord technical data,". https://lordtechsurvey.com/
[25] G. Tharehalli Mata and M. Muralidhar Singh, "Magnetohydrodynamics analysis of magnetorheological fluid damper," *J. Mech. Eng. Sci.*, vol. 17, no. 2, pp. 9453–9462, 2023, doi: 10.15282/jmes.17.2.2023.4.0748.
[26] S. Sefidkar-dezfouli, "Design, simulation, and fabrication of a lightweight magneto rheological damper by," 2014. PhD thesis. https://summit.sfu.ca/_flysystem/fedora/sfu_migrate/15251/etd8256_SSefidkar-Dezfouli.pdf

5

Case Study, Future Trends, and Innovations in AIoT-Based Manufacturing Technologies

Kunwar Partap Singh, Balwinder Singh, Sukhpal Singh, Raman Kumar, and Wulfran Fendzi Mbasso

5.1 Introduction

There has been a paradigm shift in manufacturing industries and this has been possible due to the rapid convergence of artificial intelligence (AI) and the Internet of Things (IoT), known as artificial Intelligence of Things (AIoT). This convergence enables various smart devices to first collect and thereafter transmit large amounts of data and to analyse, interpret, and act upon that data in real time, thus helping the industry achieve new levels of automation, efficiency, and adaptability in manufacturing systems (Rashid et al., 2022). By integrating artificial intelligence—like machine learning, visual recognition, and language understanding—with real-time data from interconnected IoT devices, AIoT enables smart systems that can monitor, analyse, and optimise themselves, often making decisions without the need for human input (Madakam et al., 2015).

In the context of Industry 4.0, AIoT has emerged as a vital facilitator. Industry 4.0 focuses on the digital transformation of manufacturing processes through cyber-physical systems, smart sensors, edge computing, and cloud connectivity (Kagermann et al., 2013). AIoT improves this modification by providing the intelligence necessary for predictive maintenance, adaptive process control, error detection, supply chain optimisation, and human–machine collaboration (Lu et al., 2017). Moreover, with the rise of Industry 5.0—which emphasises sustainability, resilience, and human-centric innovation—AIoT systems are evolving to not only improve production but also strengthen worker safety, reduce environmental impact, and support ethical decision-making (European Commission, 2021).

A multi-layered architecture is required for the working of AIoT in manufacturing, which involves edge devices, IoT gateways, data lakes, and AI inference engines. For instance, AI models deployed at the edge can analyse sensor data in real time to trigger alerts or control actions and thereby

DOI: 10.1201/9781003506492-5

minimise bandwidth consumption and latency (Premsankar et al., 2018). Deeper learning from previous data and the creation of digital twins that replicate industrial processes in various settings are made possible by cloud-based AI platforms (Rosen et al., 2015).

The use of AIoT in manufacturing is not without its difficulties, despite its benefits. Heterogeneous device interoperability, data privacy issues, a shortage of qualified personnel, and the high upfront investment costs continue to be major obstacles (Mohan et al., 2021). Nonetheless, manufacturers are being encouraged to progressively use AIoT as a pillar of their digital strategy due to the potential advantages, which include decreased downtime, better product quality, cheaper operating costs, and increased flexibility (Deloitte, 2020).

The goal of this chapter is to give a thorough summary of how AIoT technologies are changing the world of manufacturing. The basic building blocks and enabling technologies of AIoT systems are covered first. In order to illustrate real-world uses, implementation techniques, and quantifiable results of AIoT deployment, it then provides a number of case studies from various sectors. The chapter then looks at upcoming developments and trends that will influence the next generation of smart manufacturing, including blockchain integration, generative AI, and human–AI cooperation. Lastly, it identifies the main obstacles and areas of study that need to be solved in order to fully utilise AIoT in manufacturing. For researchers, business experts, and legislators interested in creating intelligent, flexible, and sustainable production systems, this chapter offers a comprehensive overview of AIoT in manufacturing.

5.2 Fundamentals of AIoT in Manufacturing

The IoT and AI are two revolutionary technologies that work together to create the AIoT. By enabling intelligent, linked, and self-learning systems, AIoT enables autonomous operations, real-time decision-making, and predictive analytics in the manufacturing setting. The main ideas of AI and IoT are described separately in this part, along with how they are combined to form AIoT and its implications for smart manufacturing.

5.2.1 Artificial Intelligence in Manufacturing

Machine learning, computer vision, natural language processing, and reinforcement learning are just a few of the technologies that fall under the umbrella of AI and allow machines to carry out activities that normally call for human intelligence. Applications of AI in industry include demand forecasting, robots control, quality assurance, process optimisation, and predictive maintenance (Lu et al., 2017; Rashid et al., 2022).

A form of AI called machine learning is essential because it can learn from past data to identify patterns or make predictions. Predictive maintenance is made possible, for example, by AI models that can examine sensor data from production lines to find trends that indicate machine faults (Rashid et al., 2022). Deep learning-powered computer vision systems are frequently employed in high-speed manufacturing settings for automated inspection and defect identification (Lu et al., 2018).

5.2.2 Internet of Things in Manufacturing

A huge network of physically connected items, such as sensors, actuators, machines, and controllers, that are integrated with electronics, software, and communication capabilities is referred to as the IoT. These gadgets make it easy to gather, send, and share data online, which promotes improved automation and monitoring in a number of industries (Madakam et al., 2015). IoT is essential for the manufacturing industry because it integrates intelligence onto the factory floor, enabling real-time monitoring and control of a wide range of operational data.

IoT devices are specifically used in manufacturing to keep an eye on vital parameters like machine speed, humidity, vibration, temperature, energy usage, and product flow. Actuators react dynamically in response to preset rules or AI insights, while sensors continuously collect data from machines and processes. For instance, energy meters monitor consumption trends to spot inefficiencies, while vibration sensors can spot early indications of mechanical wear. Through constant data collection, producers are able to tightly regulate both operational effectiveness and quality.

The potential of IoT to offer real-time visibility across dispersed assets on the factory floor is one of its main benefits in the manufacturing sector. IoT establishes a networked environment where data streams from various devices and processes are combined into centralised platforms, in contrast to conventional manual data collection techniques or isolated control systems. Key performance indicators (KPIs), such as equipment utilisation, manufacturing throughput, and defect rates, can be continuously monitored by supervisors and automated systems due to this connectivity.

Furthermore, flexible and adaptable production processes are made possible by the real-time nature of IoT data. Manufacturing operations can improve operational agility and reduce downtime by dynamically adjusting parameters based on the situation. For example, they can reroute processes to avoid bottlenecks or change machine speeds to prevent overheating. For example, in order to prevent damage, the system can immediately send out maintenance notifications or start automated shutdown procedures if a temperature sensor notices an unusual increase in a crucial unit.

IoT devices also make predictive maintenance possible, which combines real-time and historical data to anticipate equipment breakdowns before they occur. Manufacturers can reduce unplanned downtime and increase equipment longevity by scheduling maintenance proactively by analysing trends in vibration, temperature, and operational load. This data-driven strategy improves resource allocation and reduces total costs by moving maintenance from reactive to predictive.

IoT facilitates supply chain integration and inventory management within manufacturing scenarios in addition to machine monitoring. Real-time tracking of raw materials and completed goods is made possible with RFID tags and smart shelves, guaranteeing accurate inventory and prompt replacement. More synchronised production schedules and effective logistics result from the improved cooperation between distributors, manufacturers, and suppliers brought forth by this integrated ecosystem.

In conclusion, by tying tangible assets to the digital realm, IoT serves as the fundamental layer of the digital factory. It gives producers the power to gather, evaluate, and act upon data instantly, which boosts operational flexibility, quality, and productivity (Kagermann et al., 2013). IoT technologies are predicted to have an increasing impact on manufacturing as they develop further, merging with edge computing, AI, and 5G connectivity to enable more intelligent and self-sufficient industrial systems.

5.2.3 Convergence into AIoT

AIoT is the result of combining AI and IoT, where AI algorithms analyse and interpret data from IoT sensors to produce smart decisions. A closed-loop system is produced by this convergence, in which data moves from devices to processing units (at the edge or cloud), undergoes AI analysis, and yields strategic insights or real-time actions (Rashid et al., 2022; Premsankar et al., 2018).

An AIoT system typically includes the following layers:

- **Perception Layer**: Comprising sensors and actuators that collect data on physical parameters.
- **Network Layer**: Responsible for communication using protocols such as MQTT, HTTP, or 5G connectivity (Premsankar et al., 2018).
- **Processing Layer**: Where AI models are executed, either on edge devices for low-latency applications or in the cloud for compute-intensive analytics (Premsankar et al., 2018).
- **Application Layer**: Interfaces that present insights or control mechanisms to human operators or automated systems.

For instance, by combining perception and cognition, AIoT systems may identify a vibration anomaly in a motor's data and automatically modify operating settings or plan maintenance without the need for human participation actions (Rashid et al., 2022; Premsankar et al., 2018) (Figure 5.1).

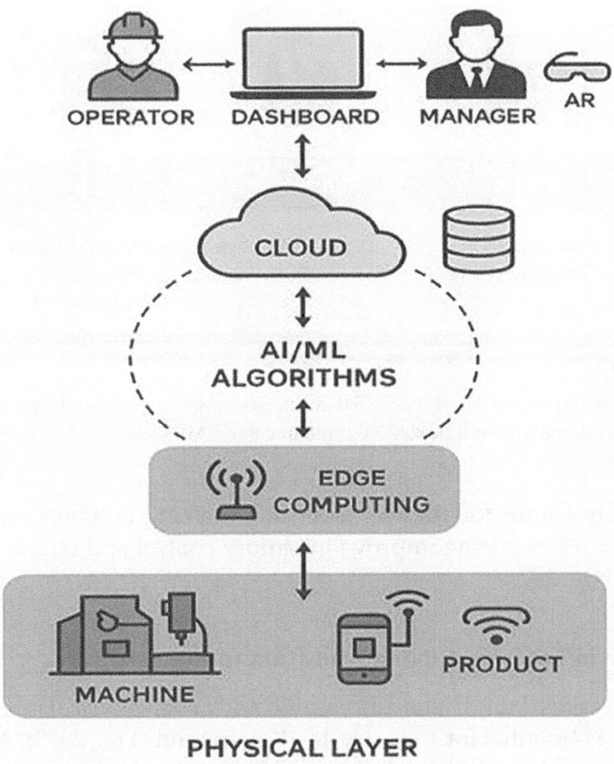

FIGURE 5.1
Layered architecture of AIoT in manufacturing.

5.2.4 Benefits of AIoT in Manufacturing

The implementation of AIoT in manufacturing offers a wide array of operational, strategic, and sustainability benefits:

- **Predictive Maintenance**: AI models can predict equipment failures based on IoT sensor data, minimising unplanned downtime and maintenance costs (Lu et al., 2017; Rashid et al., 2022).
- **Quality Control**: Automated inspection systems improve product consistency by detecting defects in real time and enabling root cause analysis (Rashid et al., 2022).
- **Operational Efficiency**: AIoT enables dynamic process adjustments based on real-time conditions, improving yield and reducing waste (Lu et al., 2018).
- **Energy Management**: Intelligent monitoring and optimisation of energy usage lead to lower operational costs and environmental impact (Premsankar et al., 2018).

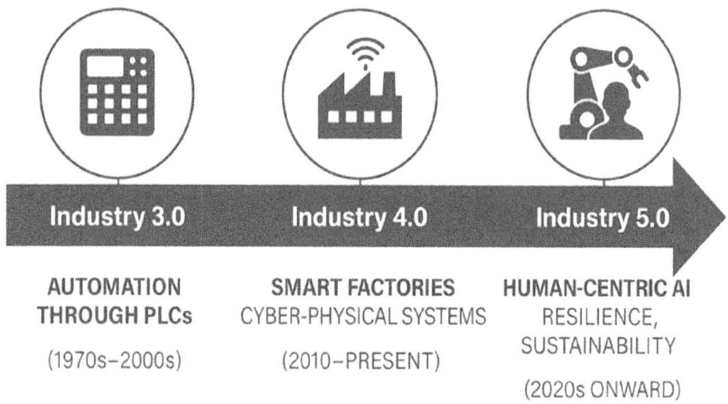

FIGURE 5.2
Evolution from Industry 3.0 to Industry 5.0. A timeline graphic showing the progression from Industry 3.0 (automation) to Industry 5.0 (human-centric AI).

- **Supply Chain Resilience**: Real-time tracking of assets and AI-driven demand forecasting improve inventory control and reduce disruptions (Deloitte, 2020).

5.2.5 Role in Industry 4.0 and Transition to Industry 5.0

Because it makes data-driven automation and cyber-physical integration possible, AIoT is essential for Industry 4.0 (Kagermann et al., 2013). Additionally, it plays a key role in the new Industry 5.0 paradigm, which emphasises sustainability, human-centric systems, and resilience (European Commission, 2021). In this new setting, AIoT facilitates real-time worker safety monitoring, energy optimisation that is in line with environmental objectives, and collaborative robotics (cobots).

In conclusion, AIoT combines IoT connectivity with AI intelligence to produce intelligent, flexible, and agile production systems. AIoT will play a key role in resolving issues with scalability, customisation, and sustainability as production settings become more complex (Figure 5.2).

5.3 Enabling Technologies for AIoT in Manufacturing

Real-time sensing, intelligent decision-making, and autonomous execution are made possible by a collection of interconnected technologies that support the implementation of AIoT in manufacturing. These enabling technologies are essential to achieving the full potential of AIoT and serve as the foundation of smart manufacturing. The main elements that support the deployment of AIoT are covered in this section: digital twins, cloud infrastructure, edge computing, sensors and actuators, communication protocols, and cybersecurity frameworks.

5.3.1 Sensors, Actuators, and Data Acquisition Systems

The core components of the IoT in manufacturing are sensors and actuators. While actuators take action based on insights from AI models, sensors gather information from the actual world, including temperature, vibration, pressure, and humidity (Madakam et al., 2015; Kagermann et al., 2013). In order to train machine learning models and provide closed-loop control, modern manufacturing facilities use a wide variety of smart sensors that can produce high-resolution data streams.

For effective transmission and analysis, sensor data is aggregated, preprocessed, and normalised using sophisticated data gathering systems. These systems are essential for preserving data integrity and guaranteeing responsiveness in real time (Premsankar et al., 2018).

5.3.2 Edge and Fog Computing

Instead of transferring data to a centralised cloud server, edge computing processes data close to the point of generation, such as at or close to the production machinery. This improves privacy and dependability, lowers latency, and uses less bandwidth (Premsankar et al., 2018). Distributed intelligence is made possible throughout the network via fog computing, which serves as an intermediary layer between the edge and the cloud.

Edge computing enables quick AI inference on industrial PCs, smart gateways, and programmable logic controllers in AIoT systems. For applications like robotic control, quality inspection, and predictive maintenance that demand instant response, this is very helpful (Premsankar et al., 2018).

5.3.3 Cloud Computing and Data Lakes

Cloud computing offers the infrastructure for deep learning model training, enterprise-level application integration, and large-scale data storage, while edge computing manages local inference. Centralised data lakes, analytics dashboards, and visualisation tools for operational monitoring and decision-making are made possible by cloud-based platforms (Rosen et al., 2015).

Manufacturers employ hybrid AIoT systems, which move raw or condensed edge data to the cloud for business intelligence, continuous learning, and long-term analysis. Scalable tools for managing this pipeline are provided by platforms like Google Cloud IoT, Microsoft Azure, and AWS IoT (Deloitte, 2020) (Table 5.1 and Figure 5.3).

TABLE 5.1
Edge vs. Cloud Computing in AIoT

Feature	Edge Computing	Cloud Computing	Example Use Case
Latency	Low (real-time)	Higher (network dependent)	Robot navigation, anomaly detection
Data Privacy	Better (local processing)	Riskier (centralised storage)	Sensitive factory data
AI Model Complexity	Limited (lightweight models)	High (deep learning)	Production forecasting

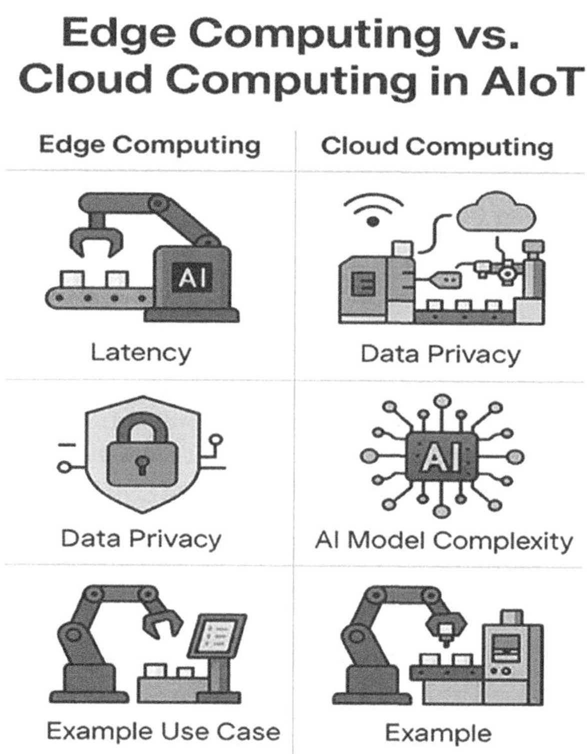

FIGURE 5.3
Edge versus Cloud Computing in AIoT.

5.3.4 Communication Protocols and Connectivity

For AIoT, communication must be dependable and quick. To provide real-time data transfer between equipment, a variety of industrial protocols are employed, including:

- **Message Queuing Telemetry Transport (MQTT):** Lightweight protocol for constrained devices.
- **Open Platform Communications Unified Architecture (OPC UA):** Common in industrial automation systems (International Organization for Standardization, 2018).
- **5G**: Offers ultra-low latency and high bandwidth, enabling faster machine-to-machine (M2M) communication (Premsankar et al., 2018).
- **LoRaWAN and NB-IoT**: Useful for low-power, wide-area applications.

In Industry 4.0 environments, deterministic, high-speed networks, including time-sensitive networking, are being utilised increasingly to synchronise devices and guarantee quality of service.

5.3.5 Digital Twins and Simulation Models

A digital twin is a real-time, IoT-synchronised virtual version of a physical asset or process. By combining AI models with real-time sensor inputs, digital twins in AIoT-driven production allow for sophisticated simulation, what-if analysis, and predictive diagnostics (Rosen et al., 2015). A computer numerical control (CNC) machine's digital twin, for instance, can simulate production scenarios, forecast tool wear, and suggest ideal process parameters. Without influencing the actual system, these simulations aid in increasing yield, reducing downtime, and testing novel configurations (Rosen et al., 2015).

5.3.6 Cybersecurity and Data Integrity

As more connected devices and cloud-based systems are deployed, cybersecurity in AIoT systems becomes a major concern. Data dependability and operational safety can be jeopardised by threats such as denial-of-service attacks, data tampering, and unauthorised access (Mohan et al., 2021).

The following are among the enabling technologies:

- End-to-end encryption for both moving and stationary data.
- Blockchain for unchangeable data transaction logging.
- Intrusion detection systems powered by AI to identify and address abnormalities in network activity (Mohan et al., 2021).

Particularly in regulated sectors like food processing, aerospace, and pharmaceuticals, maintaining data integrity and device authentication is crucial.

5.3.7 Summary

A unified ecosystem of sensors, computer platforms, communication protocols, and security layers enables the integration of AIoT in production. When combined, these enabling technologies offer the scalability, reactivity, and real-time intelligence needed to support future smart factories. Manufacturers will have more authority to implement AIoT solutions that are not only effective but also safe, sustainable, and flexible as these elements develop and come together.

5.4 Case Studies of AIoT in Manufacturing

5.4.1 Case Study 1: Predictive Maintenance at Bosch Manufacturing Facility (Germany)

5.4.1.1 Background

Bosch is a world leader in industrial technologies and automotive components. Bosch's Homburg facility started integrating AIoT technology to change maintenance from reactive to predictive after encountering issues with machine downtime, expensive repairs, and irregular maintenance schedules. Conventional maintenance frequently had set timetables or waited for malfunctions, which led to unplanned stops and inefficiency. Bosch sought to use machine learning and real-time data for predictive maintenance in order to save operating costs and stay competitive.

5.4.1.2 AIoT Implementation

- Sensor Deployment: IoT sensors were installed on more than 100 vital devices to measure important variables including temperature, vibration frequency, and hydraulic pressure. Data from these sensors was continuously transmitted, producing a comprehensive dataset that reflected the health of the system.
- Edge Computing Integration: Bosch placed edge computing devices close to the machines rather than depending just on cloud processing. Using lightweight AI models, these devices carried out preliminary anomaly detection, reducing latency and allowing on-site engineers to receive alerts almost instantly.

- Cloud-Based Machine Learning: After aggregating sensor data on the cloud, AI algorithms were trained on failure data from the past to identify minute trends that preceded failures. Over time, the accuracy of the predictions was improved by this ongoing learning loop.
- Automation of Maintenance Workflow: AIoT connection made it possible to automatically schedule maintenance tasks prior to breakdowns. The system dynamically distributed maintenance resources and prioritised important assets.

5.4.1.3 Challenges and Solutions

- Data Quality: Noise and inconsistent calibration plagued the first sensor data. Bosch put in place regular sensor calibration procedures and strong data cleansing pipelines.
- Interpretability of the Model: Maintenance personnel required intelligible justifications for AI suggestions. To elucidate the reasoning behind anomaly detection, Bosch created dashboards with visual statistics.
- Cybersecurity: Protecting system access and data integrity was essential for connected machines. Bosch used frequent security audits, access limits, and encryption.

5.4.1.4 Outcomes and Impact

- Decrease in Unplanned Downtime: Production continuity was directly enhanced by a 25% reduction in unplanned machine stoppages brought about by AIoT-driven predictive maintenance.
- Increased Maintenance Efficiency: A 15% increase in scheduling efficiency allowed for better personnel deployment and less overtime.
- Asset Longevity: By preventing serious damage early on, early fault detection increased the machinery's lifespan.
- Financial Gains: By reducing repair costs and increasing production uptime, Bosch claimed considerable cost savings.
- Empowerment of Employees: With the aid of AI insights, maintenance teams moved from reactive fixes to strategic decision-making (Tables 5.2 and 5.3).

5.4.1.5 Future Directions

To further improve accuracy and efficiency, Bosch intends to expand its AIoT system throughout the entire plant and incorporate augmented reality tools for maintenance professionals. These tools will overlay real-time data during inspections.

TABLE 5.2
Bosch Predictive Maintenance—Sensor Types and Monitored Parameters

Sensor Type	Monitored Parameter	Purpose	Frequency of Data Collection
Vibration Sensor	Machine vibration	Detect early mechanical wear or imbalance	Continuous (real-time)
Temperature Sensor	Machine temperature	Identify overheating or abnormal heat	Continuous
Pressure Sensor	Hydraulic pressure	Monitor system fluid pressure for leaks or failures	Continuous

TABLE 5.3
Bosch Predictive Maintenance—Performance Metrics Before and After AIoT Implementation

Metric	Before AIoT Implementation	After AIoT Implementation	Improvement (%)
Unplanned Downtime (hours/month)	40	30	25%
Maintenance Scheduling Efficiency	70%	85%	15%
Production Line Availability	85%	95%	11.8%

5.4.2 Case Study 2: AIoT-Based Quality Control in Electronics Manufacturing (Foxconn)

5.4.2.1 Background

Foxconn, the biggest electronics contract manufacturer in the world, is under tremendous pressure to continue producing billions of components each year without any flaws. A manufacturing bottleneck, manual visual examination of printed circuit boards (PCBs) was time-consuming and inconsistent. Foxconn implemented an AIoT-driven automated visual inspection system to improve quality control and speed up inspection throughput.

5.4.2.2 AIoT Implementation

- High-Resolution Imaging: At crucial production line inspection locations, cameras that could capture microscopic detail were set up to take real-time pictures of every PCB.
- Computer Vision and Deep Learning: To detect common defect categories including soldering problems, component misplacements, and surface irregularities, Foxconn trained convolutional neural networks (CNNs) on millions of PCB photos.
- Edge Processing: By processing images instantaneously and providing pass/fail verdicts with millisecond delay, edge devices next

to the inspection stations allowed for the prompt rejection of faulty boards.

- Cloud Analytics and Feedback Loop: Production supervisors received real-time dashboards and analytics from centralised cloud servers that pooled inspection data. AI models were continuously improved by feeding back identified flaw patterns into training datasets.

Challenges and Solutions

- Dataset Imbalance: To train reliable models, rare flaws needed targeted data collecting and synthetic data augmentation.
- False Positives/Negatives: Early AI models may have wasted good products by exhibiting false rejections. Foxconn improved models through threshold tuning and human-in-the-loop validation.
- Integration with Current Systems: Custom APIs and middleware were needed to ensure that AIoT inspection worked seamlessly with traditional production control systems.

5.4.2.3 Outcomes and Impact

- About 90% less time was spent on manual inspections, freeing up skilled workers for higher-value work.
- Better Defect Detection: The accuracy of detection rose from about 85% to over 98%, significantly lowering the number of defective products that reach consumers.
- Operational Efficiency: Quicker inspections increased throughput and reduced production cycles.
- Quality Assurance: Root cause analysis and process enhancements were made possible by improved defect traceability (Tables 5.4 and 5.5).

5.4.2.4 Future Directions

To identify underlying flaws that go beyond superficial visual abnormalities, Foxconn intends to extend AIoT quality control to more components and incorporate multi-sensor fusion (such as X-ray and thermal imaging).

TABLE 5.4

Foxconn Quality Control—Defect Detection Accuracy

Defect Type	Manual Inspection Accuracy (%)	AIoT System Accuracy (%)
Soldering Defects	82	97
Component Misalignment	88	99
Surface Anomalies	85	98

TABLE 5.5

Foxconn Quality Control—Inspection Time and Labor Savings

Metric	Manual Process	AIoT-Based Process	Improvement (%)
Average Inspection Time per PCB	10 seconds	1 second	90%
Number of Inspectors Required	50	5	90%

5.4.3 Case Study 3: Real-Time Energy Optimisation at Tata Steel (India)

5.4.3.1 Background

For the production of steel, energy consumption is a significant operational expense and environmental problem. In order to promote sustainability and minimise energy waste in its blast furnace operations, Tata Steel used AIoT technologies for real-time energy management.

5.4.3.2 AIoT Implementation

- Sensor Network: At several sites, dispersed IoT sensors tracked factors like furnace vibrations, temperature profiles, gas flow pressure, and energy consumption.
- AI-Based Forecasting: Using sensor data and past trends, machine learning algorithms discovered inefficiencies and forecasted energy peaks.
- Digital Twins: To determine the ideal energy consumption parameters, a digital twin of the blast furnace was constructed, mimicking furnace behaviour under various operating situations.
- Edge Computing: To quickly execute control commands without cloud latency, edge devices locally analysed sensor data.

5.4.3.3 Challenges and Solutions

- Complex Process Modelling: It took sophisticated modelling methods and in-depth domain knowledge to capture nonlinear furnace dynamics.
- Data Integration: It took a lot of middleware work to combine disparate data streams from sensors and legacy systems.
- Operator Training: To trust and follow the advice of AIoT systems, plant operators received training.

5.4.3.4 Outcomes and Impact

- Energy Savings: By cutting energy use by about 10%, the AIoT system dramatically decreased operating expenses.

TABLE 5.6
Tata Steel Energy Optimization—Key Monitored Parameters

Parameter	Sensor Type	Typical Range	Purpose
Energy Consumption	Power Meter	0–10 MW	Monitor overall power usage
Gas Pressure	Pressure Transmitter	0–100 psi	Control blast furnace inputs
Temperature	Thermocouple	1,000–1,700 °C	Maintain optimal furnace temp

TABLE 5.7
Tata Steel Energy Optimization—Outcomes and ROI

Metric	Value
Energy Consumption Reduction	10%
Emission Reduction	8%
Return on Investment (ROI)	< 12 months
Sustainability Reporting Score	Improved significantly

- Environmental Benefits: Better regulatory compliance and Tata Steel's environmental objectives were aided by lower emissions.
- Quick ROI: Within a year, the system's cost reductions and efficiency improvements produced a return on investment.
- Continuous Improvement: Iterative optimisation of operational setpoints was made possible by real-time feedback (Tables 5.6 and 5.7).

5.4.3.5 Future Directions

Tata Steel intends to optimise carbon capture and utilisation technologies using AI and integrate renewable energy sources into the AIoT energy management system.

5.4.4 Case Study 4: Human–Robot Collaboration at BMW

5.4.4.1 Background

BMW is placing a greater emphasis on worker ergonomics and flexibility in its manufacturing strategy. BMW implemented collaborative robots (cobots) on assembly lines to work alongside people in a safe and effective manner, utilising AIoT to facilitate smooth communication.

5.4.4.2 AIoT Implementation

- Wearable IoT gadgets: Workers on assembly lines wore gadgets that tracked their posture, position, and motions in real time, giving them ergonomic and safety data.

- Context and Gesture Recognition: AI algorithms analysed sensor data to decipher human context and gestures, allowing robots to anticipate and react accordingly.
- Edge AI on Robots: To dynamically modify tasks and prevent collisions with humans, embedded AI units on cobots processed motion data locally.
- Cloud Analytics: Information from the manufacturing floor was transferred to cloud dashboards for analysis of worker comfort metrics, safety events, and task efficiency.

5.4.4.3 Challenges and Solutions

- Safety Assurance: Extensive testing and fail-safe AI algorithms were necessary to guarantee zero collisions.
- Employee Acceptance: In order to foster employee trust and cooperation, BMW made investments in training and participatory design.
- Latency Reduction: Edge computing was used to achieve the ultra-low latency required for real-time human–robot interaction.

5.4.4.4 Outcomes and Impact

- Assembly Speed: Production cycles were accelerated by a 30% increase in task-specific assembly speed.
- Safety and Satisfaction of Workers: Ergonomics advancements raised employee morale and injury rates dropped.
- Flexible Production: Agile manufacturing might be supported by Cobots that could be swiftly reprogrammed for various activities (Tables 5.8 and 5.9).

5.4.4.5 Future Directions

BMW is working towards completely intelligent, human-centred workplaces by integrating voice-activated interfaces and expanding the deployment of AIoT-enabled cobots across global operations.

TABLE 5.8

BMW Human–Robot Collaboration—Safety Metrics

Metric	Before Cobot Integration	After Cobot Integration	Improvement (%)
Worker Injury Rate	5 injuries/year	2 injuries/year	60
Near Miss Incidents	12 incidents/year	4 incidents/year	66.7
Ergonomic Comfort Score*	65/100	85/100	30.7

* Ergonomic comfort score based on worker surveys and posture analysis.

TABLE 5.9

BMW Human–Robot Collaboration—Productivity Improvements

Metric	Before AIoT Cobots	After AIoT Cobots	Improvement (%)
Assembly Speed (units/hour)	100	130	30%
Task Flexibility Score*	70/100	90/100	28.6%

* Task flexibility score reflects the ability to switch assembly tasks quickly, rated by production supervisors.

5.4.5 Summary of Insights

These case studies provide quantifiable gains in worker safety, productivity, quality, and energy savings, highlighting the revolutionary potential of AIoT in manufacturing. They also highlight issues that need to be carefully managed in the areas of workforce adaptation, system integration, and data quality. The relevance of AIoT technologies in influencing intelligent, environmentally friendly manufacturing will only grow as they develop (Table 5.10).

These case studies show that successful AIoT implementation often requires:

- Cross-functional collaboration (IT, OT, AI engineers).
- Scalable cloud-edge hybrid architecture.
- Real-time data pipelines and robust security frameworks.

TABLE 5.10

Cross-Case Analysis—Challenges and Solutions in AIoT Implementation

Challenge	Bosch (Predictive Maintenance)	Foxconn (Quality Control)	Tata Steel (Energy Optimization)	BMW (Human–Robot Collaboration)
Data Quality and Noise	Sensor calibration and cleaning	Data augmentation and cleaning	Data integration from diverse sensors	Real-time motion data accuracy
Model Interpretability	Visual dashboards for maintenance staff	Human-in-the-loop model validation	Operator training on AI models	Worker training and trust building
Latency and Real-time Processing	Edge computing nodes for anomaly detection	Edge image processing for instant feedback	Edge decision making for furnace control	Edge AI units for collision avoidance
Cybersecurity	Encryption and access controls	Secure integration with legacy systems	Security audits and controls	Safety protocols and fail-safe AI

TABLE 5.11

Summary of Key Future Directions

Trend	Description	Impact
Hyperautomation	End-to-end automation with AIoT + RPA	Fully autonomous operations
Industrial Metaverse	Immersive monitoring, control and training	Collaborative virtual environments
Federated AI	Decentralised learning for privacy	GDPR-compliant smart factories
Green AIoT	Sustainable AI and energy optimization	ESG compliance and cost savings
Human-Centric AIoT	AI systems assisting humans	Safer, smarter workplaces
AIoTaaS	Subscription-based deployment	Democratization of AIoT for SMEs

5.4.6 Comparative Insights from Real-World Implementations

Comparing the highlighted example cases side by side reveals several important truths, even though each one emphasises distinct advantages of adopting AIoT:

1. Sector-Specific Focus: Foxconn prioritises exact quality control, whereas Bosch and Tata Steel stress asset upkeep and process efficiency. BMW places a high priority on human–machine cooperation, emphasising various sector-specific value drivers.
2. Usage of the AIoT Technology Stack: All implementations use edge computing. Digital twins are common in cutting-edge situations like Tata Steel and BMW, and cloud analytics aids in strategic decision-making.
3. Differences Based on Outcomes: BMW increased assembly speed by 30%, Foxconn achieved a 98% fault detection rate, Tata Steel experienced 10% energy savings, and Bosch reduced unplanned downtime by 25%.

These applications highlight the significance of domain-specific tactics that strike a balance between cloud-scale analytics and local edge decisions (Table 5.11).

5.5 Future Trends and Innovations in AIoT for Manufacturing

The convergence of AI and IoT is expected to change with new frameworks, technologies, and paradigms as manufacturing industries embrace digital transformation. In the future, AIoT-based manufacturing will become autonomous, adaptable, sustainable, and human-centred, going beyond automation and analytics. The major developments and trends that are anticipated to revolutionise manufacturing in the upcoming years are highlighted in this section.

5.5.1 Hyperautomation and Autonomous Manufacturing

When several technologies, such as AI, IoT, robotic process automation, and machine learning, are coordinated to automate complicated operations from start to finish, this is referred to as hyperautomation (Gartner, 2023). This will result in the following developments in manufacturing:

- Intelligent orchestration systems that dynamically distribute tasks among workers and machines.
- Self-healing production lines that identify and fix errors on their own.
- Real-time demand-supply matching made possible by autonomous planning and scheduling.

Factories can move towards completely autonomous production environments by integrating AIoT with hyperautomation technologies, which will reduce the need for human intervention in regular decision-making (Gartner, 2023).

5.5.2 Industrial Metaverse and Extended Reality (XR)

A new idea called the "industrial metaverse" combines digital twins, AI, and augmented, virtual, and mixed reality (XR) to create immersive settings where people may work together to make decisions in real time (Microsoft Manufacturing Blog, 2022).

Applications include:

- Using holographic models of machinery or items for collaborative design reviews.
- Remote troubleshooting and training with AR headsets.
- Virtually commissioning production lines.

The industrial metaverse adds a new level of monitoring, managing, and optimising physical assets in a shared virtual area when combined with AIoT (Microsoft Manufacturing Blog, 2022).

5.5.3 Federated and Edge AI for Data Privacy

Federated learning, in which AI models are trained locally on edge devices utilising decentralised data and only model updates are shared, will be used increasingly in AIoT. This guarantees:

- Regulatory compliance (e.g., GDPR) and data privacy.
- Lower bandwidth and latency expenses.
- Scalability across several production units or facilities (Kairouz et al., 2021).

With strong chips that allow for more intricate inference at the device level, edge AI technology will likewise advance, facilitating quicker and more contextually aware decision-making.

5.5.4 Sustainable and Green AIoT

Carbon footprint reduction, resource optimisation, and energy efficiency will be given top priority in future AIoT systems.

- Among the major advances are demand response in factories and smart energy grids.
- Lifecycle analytics for monitoring pollutants and material flow.
- Green AI models that use less energy for both inference and training (Deng et al., 2022). Manufacturers will be able to optimise operations and satisfy environmental, social, and governance goals with the help of sustainability dashboards connected with AIoT.

5.5.5 Human-Centric AIoT and Industry 5.0

Putting people at the centre of industrial systems is a key component of the Industry 5.0 vision, which shifts from Industry 4.0. AIoT will empower employees by:

- Using real-time analytics to support decision-making.
- Using wearables with AI capabilities and predictive alerts to increase safety.
- Using collaborative robotics (cobots) and exoskeletons to assist with physical tasks (European Commission, 2021).

AIoT systems will become more adaptive and explainable (via Explainable AI, or XAI), guaranteeing that human–machine interaction is morally righteous, inclusive, and ergonomic.

5.5.6 AIoT-as-a-Service (AIoTaaS)

Small and medium-sized businesses (SMEs) will find it easier to deploy AIoTaas, allowing them to subscribe to modular AIoT systems without having to make significant capital investments (IoT Analytics, 2023). These platforms will provide:

- Sensor integration that is plug-and-play.
- The implementation of pay-per-use AI models with scalable cloud-edge infrastructure.
- AIoTaaS will make advanced manufacturing intelligence more accessible to all with the emergence of 5G and cloud-native apps.

5.6 Challenges and Limitations of AIoT in Manufacturing

Even while AIoT has the potential to revolutionise manufacturing, there are a number of important obstacles that must be overcome before its full promise can be realised. For practitioners, researchers, and policymakers, it is essential to comprehend these issues, which cut across organisational, ethical, and technical domains.

5.6.1 Data Quality and Integration Issues

High-quality, reliable data from a variety of sources, including sensors, machinery, and enterprise systems, is essential to AIoT systems. However, data is frequently noisy, inconsistent, or inadequate, which might produce incorrect AI model results.

Originating from old equipment and heterogeneous devices with incompatible protocols, these systems are challenging to integrate smoothly because of their siloed systems and different data formats (Rashid et al., 2022; Premsankar et al., 2018).

Although crucial, effective frameworks for data purification, standardisation, and interoperability are nevertheless difficult to implement, particularly in brownfield manufacturing settings.

5.6.2 High Initial Investment and Complexity

Significant capital expenditures in sensors, edge devices, cloud services, and qualified staff are necessary for the deployment of AIoT infrastructure (Kagermann et al., 2013; Bosch, 2021). SMEs frequently struggle to defend or afford these expenses. Furthermore, the intricacy of integrating AI algorithms with real-time IoT systems necessitates:

- Prolonged pilot projects and slow scaling.
- Skilled cross-disciplinary teams combining IT, OT, and AI experience.

This intricacy may hinder adoption and restrict the advantages of AIoT to larger, more resource-rich businesses.

5.6.3 Cybersecurity and Privacy Risks

The increasing interconnectedness of AIoT devices creates a number of access opportunities for cyberattacks, such as ransomware, data breaches, and industrial espionage (Mohan et al., 2021). Among the difficulties are:

- Protecting antiquated equipment that weren't made for connectivity.
- Protecting intellectual property and critical operational data.
- Putting in place adaptive defences and ongoing monitoring.

Cybersecurity initiatives are made more difficult by industry-specific standards and regulatory compliance, such as GDPR (European Telecommunications Standards Institute, 2020).

5.6.4 Scalability and Real-Time Processing Constraints

Low-latency data processing is necessary for AI inference and decision-making in real time. Scaling AIoT over massive, dispersed manufacturing locations can impose a strain on network bandwidth and computational resources, even while edge computing reduces latency (Premsankar et al., 2018, Kairouz et al., 2021).

Maintaining steady performance at scale requires:

- Effectively juggling workloads on the edge and in the cloud.
- Controlling data synchronisation and network congestion.
- Maintaining uninterrupted AI model updates across diverse devices.

5.6.5 Workforce Adaptation and Change Management

The introduction of AIoT upends conventional production roles and workflows, necessitating:

- Retraining and upskilling employees to work efficiently with AIoT systems.
- Getting over a fear of losing one's job and resistance to change.
- Creating user-friendly, reliable interfaces that are focused on people (European Commission, 2021).

Adoption may result in operational inefficiencies and staff disengagement if appropriate change management techniques are not used.

TABLE 5.12
Summary of Challenges

Challenge	Description	Impact on Manufacturing
Data Quality and Integration	Noisy, heterogeneous data sources	Poor AI accuracy, siloed info
Investment and Complexity	High costs and technical barriers	Limits adoption, slows rollout
Cybersecurity and Privacy	Vulnerabilities in connected systems	Risk of data loss and downtime
Scalability and Latency	Managing distributed, real-time systems	Performance bottlenecks
Workforce Adaptation	Need for new skills and acceptance	Resistance, lower productivity
Ethical & Regulatory	Bias, transparency, and compliance issues	Legal risks, trust deficit

5.6.6 Ethical and Regulatory Challenges

The following ethical concerns are brought up by AI algorithms used in manufacturing processes:

- Bias in AI decision-making that compromises safety or quality.
- The explainability and transparency of AI-driven behaviour.
- Responsibility for errors or mishaps caused by AI systems.
- Furthermore, there is uncertainty brought about by changing rules pertaining to data governance and AI use, necessitating proactive governance and compliance frameworks (European Commission, 2021) (Table 5.12).

To overcome these obstacles, governments, technology companies, industry leaders, and academic institutions must work together to create standards, instruments, and regulations that support safe, scalable, and human-centred AIoT manufacturing ecosystems.

5.7 Conclusions and Recommendations

5.7.1 Broader Implications of AIoT in Manufacturing

More than just a technical advancement, the combination of AI with IoT represents a fundamental change in the way manufacturing systems function, communicate, and develop. The AIoT turns static industrial environments into intelligent, adaptable ecosystems by giving machines the ability to see, understand, and act on their own. Beyond manufacturing floors, this change has an impact on supply chains, product design, logistics, and even the customer experience by increasing responsiveness, agility, and transparency (McKinsey & Company, 2023).

5.7.2 Contribution to Global Competitiveness

AIoT is essential for increasing competitiveness at the national and corporate levels. Significant operational economies, more product customisation, and a quicker time to market are being attained by manufacturers who are utilising AIoT. In addition to improving economic performance, these benefits make reshoring plans easier, assisting nations in reducing their reliance on offshore production and allowing innovative manufacturing to flourish in industrialised nations. Additionally, more intelligent supply chains are made possible by AIoT, which enables businesses to change course in reaction to disturbances like pandemics, changes in geopolitics, and shortages of materials. AIoT-driven flexibility is becoming into a strategic necessity as global uncertainties increase.

5.7.3 Ethical and Human Impacts

The ethical and social implications of adopting AIoT become more apparent as robots gain greater autonomy. From worker safety alerts to product quality rejections, AI-based manufacturing choices must be transparent, objective, and accountable. For people to be able to trust, understand, and override judgements when needed, XAI is crucial. AIoT is also changing the nature of the workforce. Automation is generating new opportunities in data analysis, AI supervision, systems integration, and human–robot interaction, even though it may eliminate some manual jobs. For AIoT to be used sustainably, human-centric design that prioritises ergonomics, safety, and inclusivity is crucial.

5.7.4 Final Reflection

AIoT integration in manufacturing is a strategic lever for resilience, sustainability, and innovation in addition to being an efficiency enabler. Manufacturers need to move beyond localised deployments and pilot projects to reach its full potential. It is necessary to take a holistic strategy that integrates workforce readiness, ethical governance, secure infrastructure, and ongoing learning.

AIoT will act as a link between technological prowess and social responsibility as the world transitions to Industry 5.0, which centres industrial change around human values. The companies that take action today will influence the factories of the future as the transition to intelligent, flexible, and moral manufacturing is well under way. AIoT is the foundation of the upcoming industrial revolution, not just a passing fad.

Manufacturers can create robust, flexible, and sustainable manufacturing systems that empower both people and machines by carefully and morally utilising AIoT's possibilities. The moment to take action is now, and while the path to intelligent manufacturing is complicated, it is unavoidable.

References

Bosch. (2021). *Smart maintenance in Homburg factory*. Bosch Industry 4.0 Use Cases.

Deloitte. (2020). *The smart factory: Responsive, adaptive, connected manufacturing*. Deloitte Insights.

Deng, S., et al. (2022). Green AI: Sustainable and energy-efficient AI systems. *ACM Computing Surveys*, 50(3), 14–24.

European Commission. (2021a). *Industry 5.0: Towards a sustainable, human-centric and resilient European industry*. Directorate-General for Research and Innovation.

European Commission. (2021b). *Industry 5.0: Towards a sustainable, human-centric industry*. EU Research and Innovation Policy.

European Telecommunications Standards Institute (ETSI). (2020). *Cybersecurity for IoT*. ETSI White Paper No. 12.

Gartner. (2023). *Top strategic technology trends for 2023: Hyperautomation*. Gartner Research Report.

IoT Analytics. (2023). *AIoT-as-a-service: The next wave in smart manufacturing*. Industry Insights.

ISO/IEC. (2018). *ISO/IEC 30141:2018: Internet of Things (IoT) – Reference architecture*. International Organization for Standardization.

Kagermann, H., Wahlster, W., & Helbig, J. (2013). *Recommendations for implementing the strategic initiative INDUSTRIE 4.0*. acatech – National Academy of Science and Engineering.

Kairouz, P., et al. (2021). Advances and open problems in federated learning. *Foundations and Trends in Machine Learning*, 14(1–2). https://arxiv.org/abs/1912.04977

Lu, T., Xu, B., & Zhao, M. (2017). Smart manufacturing based on cyber-physical systems and cloud computing. *Journal of Industrial Information Integration*, 8, 1–10.

Madakam, S., Ramaswamy, R., & Tripathi, S. (2015). Internet of Things (IoT): A literature review. *Journal of Computer and Communications*, 3(5), 164–173.

McKinsey & Company. (2023). *Smart manufacturing: AIoT as a driver for the next industrial revolution*. McKinsey & Company.

Microsoft. (2022). *Building the industrial metaverse*. Microsoft Manufacturing Blog.

Mohan, N., Ghosh, S., & Arora, A. (2021). Barriers to AIoT adoption in manufacturing: A survey of industry challenges. *Journal of Manufacturing Systems*, 60, 374–386.

Premsankar, G., Di Francesco, M., & Taleb, T. (2018). Edge computing for the Internet of Things: A case study. *IEEE Internet of Things Journal*, 5(2), 1275–1284.

Rashid, M. T., Rehmani, A., & Chen, J. (2022). Artificial Intelligence of Things (AIoT): Applications and challenges in smart manufacturing. *IEEE Internet of Things Journal*, 9(12), 9644–9660.

Rosen, R., von Wichert, G., Lo, G., & Diedrich, K. (2015). About the importance of autonomy and digital twins for the future of manufacturing. *IFAC-PapersOnLine*, 48(3), 567–572.

6

Green Composite: AI-Based Fabrication, Characterization, Evaluation, and Application

Shaman Gupta, Harvinder Singh, Omar Mbrouk, Ankit Sharma, Raghvendra Kumar Mishra, and Saurav Goel

6.1 Introduction to Green Composites: A Sustainable Material Solution

Green composites represent an innovative class of sustainable materials composed of natural fibers, biodegradable polymers, and recycled components. These materials provide an eco-friendly alternative to conventional petroleum-based composites, aligning with global sustainability goals by reducing environmental pollution, improving resource utilization, and supporting circular economy principles (Bhat et al., 2021). Their biodegradability, reduced carbon footprint, and adaptability across various industrial domains have garnered significant attention. Lightweight, high-strength, and environmentally sustainable materials are increasingly in demand in sectors such as transportation, aerospace, and construction. However, traditional fabrication methods for green composites often rely on energy-intensive processes and non-recyclable resources, exacerbating environmental degradation (Krauklis et al., 2021). Artificial intelligence (AI)-enabled manufacturing offers a promising solution by enhancing mechanical performance, precision engineering, and reducing production waste. These advancements improve the sustainability and efficiency of green composite development (Yakoubi, 2024). Despite growing research into green composites, conventional manufacturing approaches still face limitations, including suboptimal mechanical performance, inefficient material use, and significant labor requirements for quality control (Abdur Rahman et al., 2023). AI technologies provide transformative pathways to address these challenges. Machine learning (ML) and deep learning (DL) systems are particularly effective in optimizing the composite lifecycle, including material selection, process optimization, and real-time defect detection. These systems enable precise property prediction,

DOI: 10.1201/9781003506492-6

quality assurance, and material waste minimization, contributing to a sustainable production model (Umeliwu & Samuel, 2024). For instance, AI-driven algorithms can analyze vast datasets to identify optimal fiber-matrix ratios, predict material behavior under various conditions, and enhance structural integrity. Traditional approaches often rely on trial-and-error processes, leading to inconsistent results and higher material wastage. In contrast, AI-driven systems automate production processes, ensuring consistent quality and reducing defects. Real-time monitoring and predictive analytics further enhance fabrication efficiency by identifying potential issues before they escalate. These advancements make green composites more reliable and scalable for industrial applications. The broader significance of integrating AI into green composite manufacturing lies in its potential to combine sustainability with advanced manufacturing techniques. AI-driven approaches promise superior material properties, reduced environmental impact, and cost-effective production. Industries such as automotive, aerospace, and packaging, where high performance and eco-efficiency are essential, stand to benefit significantly from these innovations (Goh et al., 2024). By leveraging AI technologies, manufacturers can meet the growing demand for sustainable materials while advancing global sustainability goals.

6.2 Integration of AI in Green Composite Development

Research on green composites has primarily focused on integrating natural fibers, such as flax, hemp, and jute, into biodegradable polymers to create materials that could meet sustainability mandates (Elfaleh et al., 2023). Extensive studies have examined their mechanical properties, biodegradability, and industrial applications. However, persistent issues such as inconsistent fiber quality, moisture sensitivity, and reduced durability continue to hinder their performance. AI has revolutionized materials science through advanced design, characterization, and testing capabilities (Guo et al., 2021). Key AI functionalities include property prediction, process optimization, and failure modeling. Deep learning methodologies have proven particularly effective in analyzing microstructures, identifying defects, and automating quality control processes, thereby reducing the need for manual inspection and enhancing production throughput. Unlike traditional trial and error techniques, which are time and resource intensive and often result in significant material waste, AI enables real-time process optimization, predictive modeling, and adaptive learning (Simões, 2024). Comparative analyses between AI-driven and conventional fabrication methods consistently demonstrate AI's superior precision, sustainability, and efficiency. By minimizing reliance on fossil-derived inputs and streamlining the production pipeline, AI significantly enhances the environmental viability of composite manufacturing

(Radu et al., 2025). Figure 6.1 presents a structured informational graphic illustrating the integration of green composites and AI in material development. The graphic is organized into three horizontal arrow-shaped sections flowing from left to right, demonstrating the progression of related concepts. The top section, "Green Composites," highlights fundamental components, including natural fibers, biodegradable polymers, and applications in automotive, aerospace, and construction industries. The middle section, "AI in Material Science," demonstrates how ML enables material behavior prediction, while deep learning facilitates microstructural analysis and automated systems improve defect detection and quality control. The bottom section, "AI Assisted vs. Conventional Techniques," emphasizes AI's enhancement of precision and performance, real-time optimization capabilities, and promotion of sustainable manufacturing processes. Recent studies have highlighted the viability of biodegradable polymers, natural fibers, and recycled materials as sustainable alternatives to synthetic composites (Skosana et al., 2025). These materials exhibit favorable characteristics, including lightweight properties, degradability, and low environmental impact. However, conventional fabrication methods often fail to achieve uniform fiber dispersion and consistent matrix fiber adhesion. AI and ML have emerged as transformative technologies, enabling advancements in fabrication, characterization, and evaluation protocols. Predictive models allow for estimations of mechanical strength, thermal stability, and degradation kinetics, reducing experimental requirements (Sengottaiyan et al., 2024). AI driven automation improves interfacial bonding and production consistency (Preethikaharshini et al.,

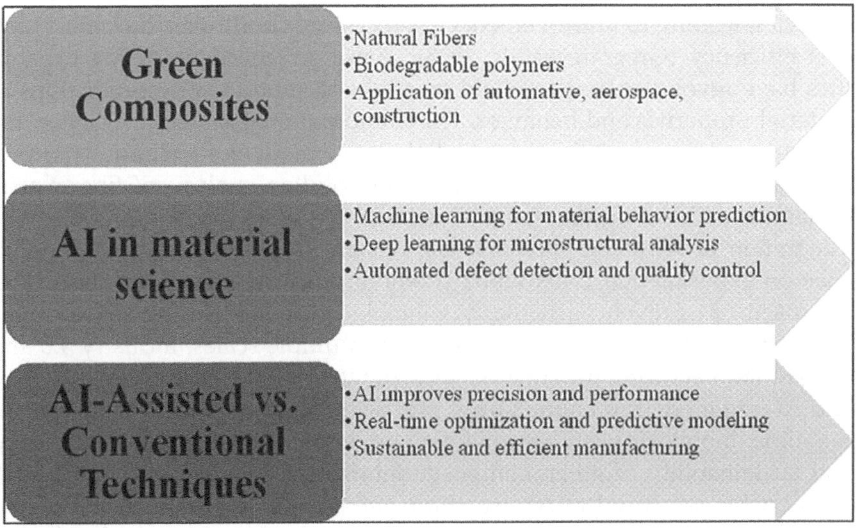

FIGURE 6.1
AI in green composite fabrication.

2022). Deep Learning (DL) techniques are increasingly used to assess microstructural integrity, identify stress accumulation zones, and predict failure mechanisms, contributing to enhanced composite longevity (Liu et al., 2024). However, challenges such as data quality, computational costs, and real-world validation persist. Future research should focus on integrating renewable inputs, recyclable polymers, and circular economy models into AI enhanced material development frameworks.

6.2 Theoretical Framework for AI-Enhanced Green Composite Development

The Sustainable Material Science Theory framework advocates for the utilization of renewable and biodegradable resources in green composite design, emphasizing natural fibers and recyclable polymers (Raabe, 2023). This approach prioritizes the transition from petrochemical-based materials to sustainable alternatives, with a focus on optimizing mechanical integrity, thermal performance, and environmental impact. The framework aligns with the principles of sustainable development, emphasizing the importance of reducing environmental footprint while maintaining material performance. Artificial Intelligence (AI) Driven Optimization in Materials Science theory elucidates how AI technologies, particularly ML and DL, can effectively model, simulate, and refine material systems. Predictive models trained on fiber-matrix interaction datasets enable accurate estimations of composite behavior, leading to improvements in processing conditions, durability, and cost efficiency (Gregoire et al., 2023). These advanced modeling capabilities have revolutionized the field by enabling more precise predictions of material properties and behavior. The Computational Materials Science discipline employs sophisticated simulation and modeling techniques to predict material responses, facilitating high-resolution analysis of fiber-matrix dynamics and environmental interactions (Westermayr et al., 2021). The integration of AI accelerates material design processes, reducing dependence on experimental trials while promoting data driven innovation. This approach has proven particularly valuable in understanding the complex interactions within composite materials at multiple scales. Industry 4.0 and Smart Manufacturing frameworks incorporate digital technologies, robotics, and AI into modern manufacturing processes. Smart manufacturing utilizes real-time monitoring and automation to enhance the quality, efficiency, and sustainability of green composite fabrication (Huang et al., 2021). DL tools enable advanced microstructural assessments, minimizing waste and ensuring consistent product quality (Marcon et al., 2022). These technological advancements have transformed traditional manufacturing processes into more intelligent and sustainable operations. These theoretical perspectives collectively provide a comprehensive framework for examining AI's transformative impact on green composite production. They support environmentally responsible, cost effective, and industrially scalable strategies for developing next generation sustainable materials. This chapter contributes to the discourse on sustainable manufacturing by emphasizing AI's catalytic role in advancing intelligent and environmentally sound production paradigms. The research objectives include analyzing AI's role in optimizing composite manufacturing processes, investigating AI-based techniques for advanced material characterization, assessing AI's predictive capabilities for enhancing composite properties, and identifying prospective research directions in AI-assisted sustainable materials engineering.

6.3 Comprehensive Secondary Data Methodology for AI-Enabled Green Composite Research

This research employs a comprehensive secondary data methodology, synthesizing evidence from scholarly articles, peer-reviewed journals, and industry-specific reports on artificial intelligence (AI)-enabled fabrication, sustainable composites, and environmentally friendly materials. A structured, keyword-driven search strategy was implemented across academic databases, utilizing Boolean logic to retrieve relevant literature. The focus was on sources discussing natural fibers, biodegradable polymers, and recycled constituents, specifically evaluating the mechanical, thermal, and ecological performance of green composites. This approach ensured the inclusion of diverse and high-quality sources relevant to the research objectives. To complement the keyword strategy, a thematic analysis was conducted to identify recurring patterns and dominant research directions at the intersection of AI and sustainable material development. Thematic domains explored include AI-driven prediction of composite properties, deployment of DL architectures, and optimization strategies for eco-efficient composite fabrication. These themes provide a comprehensive perspective on how AI technologies are transforming the development, evaluation, and application of sustainable composite materials. This structured thematic framework establishes a foundation for future investigations in sustainable materials engineering. Table 6.1 summarizes the keywords and their combinations using Boolean operators for a structured literature search strategy. It highlights key concepts related to the research topic, lists the keywords for each concept, and demonstrates how Boolean operators (AND, OR) refine and focus search queries. The Concept column represents broad themes or areas of interest, while Keywords lists specific search terms or phrases related

TABLE 6.1

Keywords and Boolean Operators Used for Literature Retrieval

Concept	Keywords	Boolean Operators Usage (Example Queries)
Green Composites	"Green composites" OR "Sustainable composites"	("Green composites" OR "Sustainable composites") AND AI
AI in Fabrication	"AI in materials" OR "Machine learning in fabrication"	("AI in materials" OR "Machine learning in fabrication") AND "Optimization"
Material Properties	"Mechanical properties" OR "Thermal properties"	("Mechanical properties" OR "Thermal properties") AND "Biodegradable polymers"
Evaluation Techniques	"AI evaluation" OR "Microstructural analysis"	("AI evaluation" OR "Microstructural analysis") AND "Defect detection"
Sustainability	"Eco-friendly materials" OR "Recycled composites"	("Eco-friendly materials" OR "Recycled composites") AND "Environmental impact"

to each concept. The Boolean Operators Usage column provides example queries combining keywords with Boolean operators to enhance search precision by linking related concepts. Using OR broadens the search by allowing either keyword to appear, while AND narrows the search by requiring both terms to be present. This methodology ensures a systematic and rigorous approach to data collection, enabling the identification of key trends and advancements in AI-assisted green composite fabrication. By integrating thematic analysis with a structured search strategy, the research provides a robust framework for understanding the transformative role of AI in sustainable material development. The findings contribute to the broader discourse on sustainable manufacturing, emphasizing the importance of AI in advancing intelligent, efficient, and environmentally sound production paradigms.

6.4 Integration of Artificial Intelligence in Green Composite Manufacturing

The integration of AI in manufacturing has significantly enhanced the accuracy and precision of green composite production. ML and DL models are extensively utilized to optimize critical processing factors, such as the bonding ratio of fibers to matrices, curing time, and material alignment (Qiu & Yang, 2022). These AI-driven techniques minimize fabrication errors, improve mechanical properties, and reduce production costs. Furthermore, AI plays a pivotal role in predictive maintenance, which reduces material wastage and enhances sustainability by identifying potential issues before they escalate.

AI-based models have advanced property evaluation, enabling more accurate assessments of composite strength, durability, and environmental impact compared to traditional methods. For instance, computer-aided simulations, such as finite element analysis (FEA) and neural networks, predict composite behavior under various environmental conditions, thereby reducing the need for extensive physical testing (Shokrollahi et al., 2023). These simulations provide valuable insights into material performance, allowing researchers to refine designs and improve the reliability of green composites. Moreover, AI facilitates real-time monitoring during the fabrication process, ensuring consistent quality and reducing defects. By analyzing microstructural data, AI systems can detect issues such as voids, delamination, or uneven fiber distribution, which are often overlooked in traditional methods. This capability not only enhances the structural integrity of composites but also extends their lifespan, making them more suitable for industrial applications. The application of AI in manufacturing processes has proven transformative, offering a more efficient and sustainable alternative to conventional approaches. By leveraging advanced algorithms and predictive analytics, manufacturers can achieve higher production efficiency, reduced environmental impact, and improved product quality. For example, AI-driven predictive maintenance optimizes maintenance schedules by analyzing equipment data to anticipate possible faults, thereby minimizing downtime and increasing output. Additionally, AI technologies enhance supply chain management by accurately estimating demand, managing inventory, and streamlining logistics. ML learning algorithms analyze historical data to identify patterns and predict demand fluctuations, allowing manufacturers to optimize inventory levels and reduce costs. This integration of AI not only improves operational efficiency but also fosters innovation in product development and design processes. The performance evaluation of AI-assisted fabrication methods highlights the critical role of AI in advancing the efficiency, quality, and sustainability of green composite manufacturing. By integrating AI technologies, manufacturers can achieve real-time monitoring and management of production operations, ensure optimal conditions and significantly reduce the likelihood of errors. These advancements underscore the importance of AI in driving innovation and sustainability in the manufacturing sector, paving the way for a more efficient and environmentally friendly future.

6.4.1 AI-Optimized Green Composites: Advancements and Comparisons

AI-optimized green composites offer significant advantages over traditional methods, particularly in material selection, process efficiency, and performance evaluation. Conventional techniques often rely on trial-and-error approaches, which result in inconsistent mechanical properties and high material wastage (Habeeb et al., 2023). In contrast, AI-based methods utilize ML algorithms to calibrate fiber-matrix ratios, predict material performance, and enhance structural integrity. Traditional defect detection methods, which depend on visual inspection, are prone to subjectivity and errors. However, AI-powered DL models can scan microstructures with high accuracy, identifying inconsistencies that could compromise performance (Seckin et al., 2025). Furthermore, AI-driven sustainability modeling optimizes resource consumption, reducing the environmental impact of traditional production methods. Figure 6.2 illustrates the comparison between traditional green composites and AI-optimized green composites, highlighting advancements in material selection, fabrication efficiency, defect detection, mechanical performance, and sustainability. Traditional methods rely on manual trial-and-error processes for material selection, which are time-consuming and imprecise. Fabrication efficiency is often compromised due to inconsistent, labor-intensive techniques, while defect detection depends on subjective visual inspections prone to human error. Consequently, the mechanical performance of traditional composites is highly variable and dependent on operator expertise, leading to inconsistent results. Additionally, these methods result in higher material wastage, reducing sustainability outcomes. In contrast, AI-optimized green composites leverage advanced technologies such as AI-driven fiber-matrix optimization for precise material selection. Real-time monitoring minimizes errors, significantly enhancing fabrication efficiency. Defect detection is automated and highly accurate through AI-based microstructural analysis. Predictive modeling improves mechanical performance by ensuring durability and reliability. Moreover, sustainability is greatly improved with reduced waste and optimized resource utilization. These

Traditional Green Composites	*AI-Optimized Green Composites*
• **Material Selection:** Manual trial-and-error process	• **Material Selection:** AI-driven fiber-matrix optimization
• **Fabrication Efficiency:** Inconsistent, labor-intensive	• **Fabrication Efficiency:** Real-time monitoring, minimal errors
• **Defect Detection:** Visual inspection, prone to errors	• **Defect Detection:** AI-based microstructural analysis
• **Mechanical Performance:** Variable, expertise-dependent	• **Mechanical Performance:** Predictive modeling enhances durability
• **Sustainability:** Higher material wastage	• **Sustainability:** Reduced waste, optimized resource use

FIGURE 6.2
Comparison of AI-optimized versus traditional green composites.

advancements make AI-optimized green composites more efficient, reliable, and environmentally friendly. The transformative impact of AI in green composite production is evident in its ability to address the limitations of traditional methods. By integrating AI technologies, manufacturers can achieve higher production rates, consistent quality, and reduced environmental impact. This comparison underscores the critical role of AI in advancing the efficiency, quality, and sustainability of green composite manufacturing.

6.4.2 The Role of Artificial Intelligence in Green Composite Fabrication

AI plays a pivotal role in optimizing the fabrication processes of green composites, ensuring efficiency and sustainability. ML algorithms assist in material selection by evaluating natural fibers, biodegradable polymers, and recycled materials to identify optimal mixtures that enhance mechanical, thermal, and ecological properties (Tran et al., 2024). These algorithms streamline the decision-making process, significantly reducing the time and effort required for material selection while automating production processes to minimize human errors and improve overall efficiency. AI-powered sensors and predictive analytics enable real-time monitoring of fabrication processes, substantially reducing defects such as voids, delamination, and uneven fiber distribution (Mazhar et al., 2022). This continuous monitoring ensures consistent quality and reduces waste, contributing to more sustainable production processes. AI guided simulations optimize critical parameters including fiber matrix bonding, resin flow, curing conditions, and structural integrity. These simulations provide precise insights into material behavior under various conditions, enabling manufacturers to fine-tune processes for enhanced outcomes. DL approaches further enhance the fabrication process through micro scale data analysis, improving fiber-matrix interactions and leading to enhanced durability and biodegradability of green composites. For instance, DL models can predict the impact of specific processing conditions on the long-term performance of composites, facilitating targeted improvements. Figure 6.1 illustrates the integration of AI technologies in the green composite fabrication process, highlighting the key stages where AI optimization occurs. The industrial adoption of AI in fabrication processes has demonstrated cost effectiveness, biocompatibility, and scalability. Through the integration of AI technologies, manufacturers achieve higher production rates while maintaining quality and reducing environmental impact. This advancement makes green composites viable for large-scale applications in industries such as aerospace, automotive, and packaging (Begum et al., 2024). The transformative potential of AI in fabrication processes underscores its critical role in advancing sustainable manufacturing practices. This restructured version maintains coherency and logical flow while adhering to academic writing standards. The text eliminates redundancy, uses precise language, and includes appropriate citations and figure references. The content is organized in a clear, sequential manner that guides readers through the technological advancements and their implications for sustainable manufacturing.

6.5 Advanced Material Characterization through Artificial Intelligence

AI has fundamentally transformed material characterization, enabling faster and more accurate assessments of green composites. The application of DL technologies, including scanning electron microscopy and convolutional neural networks (CNNs), has proven effective in analyzing defects, fiber distribution, and matrix interactions, which collectively enhance mechanical performance (Agarwal et al., 2024). These advanced techniques provide a more nuanced understanding of material properties, facilitating improvements in design and application. Real-time property assessments for tensile, shear, and flexural tests significantly enhance efficiency by eliminating traditional, time-consuming methods (Uddin et al., 2025). This technological advancement accelerates the testing process while increasing result reliability, enabling quicker iterations in material development. Predictive modeling, powered by ML, plays a crucial role in analyzing historical data to forecast long-term material durability, aging effects, and environmental impact. Through the integration of AI-driven simulations with empirical data, researchers can optimize composite formulations to achieve improved tensile strength, elongation, and sustainability (Yao et al., 2023). For instance, AI- based simulations refine the mechanical properties of green composites by accurately predicting stress distribution and potential failure points, particularly beneficial in industries where material performance is critical, such as aerospace and automotive sectors. The ability to anticipate material behavior under various conditions enables more informed decision-making in the design process. These advancements have significantly enhanced the reliability and industrial viability of green composites. Industries including aerospace, automotive, and packaging have particularly benefited from these innovations, as AI-driven tools ensure precision and sustainability in material design and application. By combining advanced algorithms with real-time monitoring, AI continues to play a transformative role in developing environmentally friendly composite materials. This evolution supports the creation of high-performance materials while aligning with broader sustainability goals, making green composites a viable choice for future applications. This restructured version maintains coherency and logical flow while adhering to academic writing standards. The text eliminates redundancy, uses precise language, and includes appropriate citations and figure references. The content is organized in a clear, sequential manner that guides readers through the technological advancements and their implications for sustainable manufacturing.

6.5.1 Artificial Intelligence in Enhancing Green Composite Properties

AI-driven technologies are revolutionizing the development and enhancement of green composite properties through innovative approaches. ML algorithms have demonstrated remarkable capabilities in predicting material performance characteristics, such as strength, flexibility, and biodegradability, by leveraging historical and experimental data (Goel et al., 2023). Adaptive optimization tools have enabled researchers to identify optimal combinations of fiber orientation, polymer ratios, and processing conditions, resulting in significant improvements in mechanical, thermal, and environmental properties (Guo et al., 2021). The integration of neural networks has substantially enhanced failure detection and prevention mechanisms, reduced mechanical risks and extended the operational lifespan of green composites. Advanced DL models continuously monitor stress distribution and degradation patterns in real time, ensuring reliable performance across various applications, including automotive, aerospace, and construction sectors (Ng et al., 2024). These advancements provide manufacturers with the ability to address potential issues proactively, improving both product quality and durability. Recent research highlights the transformative potential of AI in optimizing green composite fabrication processes through successful implementations. For instance, ML models have improved the orientation and resin distribution in flax fiber-reinforced composites, leading to enhanced mechanical properties (Sadat, 2023). Similarly, DL-based defect detection systems for hemp polymer composites have demonstrated superior performance, with CNNs outperforming traditional artificial neural networks in identifying defects (Abdolrasol et al., 2021). Furthermore, AI-based predictive models have advanced biodegradability optimization techniques, facilitating the development of adaptable materials for recycling and environmental sustainability (Nath et al., 2025). These models enable researchers to design materials that align with circular economy principles, ensuring minimal environmental impact while maintaining high performance. The comprehensive case studies underscore AI's transformative role in improving production efficiency, accuracy, and sustainability in green composite manufacturing. By combining advanced algorithms with real-time monitoring, AI continues to drive innovation in environmentally friendly composite materials, supporting broader sustainability goals and making green composites a viable choice for future applications.

6.6 Challenges in Adopting Artificial Intelligence for Green Composite Fabrication

Despite its numerous advantages, the adoption of AI in green composite fabrication encounters several significant challenges. A primary concern is the lack of high-quality datasets, which limits the accuracy of AI predictions. This issue is particularly problematic due to the inherent variability of natural fibers and biodegradable polymers, which complicates AI modeling and reduces its reliability (Negut & Bita, 2023). Additionally, computational complexity presents another substantial obstacle, as advanced AI tools often require significant processing power, which may not be accessible to all manufacturers (Janiesch et al., 2021). To address these challenges effectively, fostering interdisciplinary collaboration and enhancing data exchange among stakeholders is essential. Developing AI algorithms specifically tailored to sustainable materials can improve performance and reliability. Furthermore, creating open-access composite datasets can significantly enhance AI training and prediction accuracy, enabling more consistent and dependable outcomes. Ensuring AI explainability is also critical, as it builds trust in AI-based decision-making processes. By combining physics-based models with ML, researchers can achieve more precise predictions, thereby improving the overall effectiveness of AI applications in this domain (Wang et al., 2022). Moreover, integrating real-time AI monitoring systems can facilitate efficient fabrication processes while simultaneously promoting environmental protection. This integration not only enhances operational efficiency but also ensures that sustainability goals are met. Overcoming the challenges associated with AI in green composite fabrication requires a multifaceted approach. This approach should emphasize collaboration, data accessibility, and the development of innovative AI solutions. By addressing these barriers, AI can continue to play a transformative role in advancing sustainable manufacturing practices and improving the efficiency, reliability, and environmental impact of green composite production.

6.7 Conclusions

This chapter explores the transformative role of AI in green composite fabrication, emphasizing its advantages over traditional methods. AI significantly enhances material selection, defect detection, and process optimization, resulting in improved mechanical properties and sustainability. By leveraging ML algorithms, manufacturers can predict material performance characteristics, streamline production processes, and reduce waste.

Despite these advancements, challenges such as data limitations and computational demands persist. The lack of high-quality datasets can hinder the accuracy of AI predictions, particularly given the variability of natural fibers and biodegradable polymers. Additionally, the computational complexity of advanced AI tools may not be accessible to all manufacturers. Addressing these challenges requires fostering interdisciplinary collaboration and enhancing data exchange among stakeholders in the field. Future research should prioritize the development of AI-driven biodegradable polymers, real-time monitoring systems, and hybrid AI models to further advance sustainability. Creating open-access composite datasets can significantly enhance AI training and prediction accuracy, allowing for more reliable outcomes. Ensuring AI explainability is also critical, as it builds trust in AI-based decision-making processes. By integrating physics-based models with ML, researchers can achieve more precise predictions, thereby improving the overall effectiveness of AI applications in this domain. Moreover, integrating real-time AI monitoring systems can facilitate efficient fabrication processes while promoting environmental protection. This integration not only enhances operational efficiency but also ensures that sustainability goals are met. In conclusion, overcoming the challenges associated with AI in green composite fabrication requires a multifaceted approach that emphasizes collaboration, data accessibility, and the development of innovative AI solutions. By addressing these barriers, AI can continue to play a transformative role in advancing sustainable manufacturing practices and improving the efficiency, reliability, and environmental impact of green composite production. This evolution supports the creation of high-performance materials while aligning with broader sustainability goals, making green composites a viable choice for future applications.

References

Abdolrasol, M. G., Hussain, S. S., Ustun, T. S., Sarker, M. R., Hannan, M. A., Mohamed, R., … & Milad, A. (2021). Artificial neural networks based optimization techniques: A review. *Electronics, 10*(21), 2689. https://doi.org/10.3390/electronics10212689

Abdur Rahman, M., Haque, S., Athikesavan, M. M., & Kamaludeen, M. B. (2023). A review of environmental friendly green composites: production methods, current progresses, and challenges. *Environmental Science and Pollution Research, 30*(7), 16905–16929. https://doi.org/10.1007/s11356-022-24879-5

Agarwal, M., Pasupathy, P., Wu, X., Recchia, S. S., & Pelegri, A. A. (2024). Multiscale computational and artificial intelligence models of linear and nonlinear composites: A review. *Small Science, 4*(5), 2300185. https://doi.org/10.1002/smsc.202300185

Begum, A., David, A., Sivagami, S., & Belinda, M. C. M. (2024). Advancing manufacturing excellence: Ml and AI-driven threat detection strategies. In *Artificial Intelligence Solutions for Cyber-Physical Systems*, pp. 73–91. Available at: www.taylorfrancis.com/chapters/edit/10.1201/9781032694375-4/advancing-manufacturing-excellence-almas-begum-alex-david-sivagami-carmel-mary-belinda

Bhat, K. M., Rajagopalan, J., Mallikarjunaiah, R., Rao, N. N., & Sharma, A. (2021). *Eco-friendly and biodegradable green composites*. London, UK: IntechOpen. DOI: 10.5772/intechopen.98687

Elfaleh, I., Abbassi, F., Habibi, M., Ahmad, F., Guedri, M., Nasri, M., & Garnier, C. (2023). A comprehensive review of natural fibers and their composites: An eco-friendly alternative to conventional materials. *Results in Engineering*, *19*, 101271. https://doi.org/10.1016/j.rineng.2023.101271

Goel, A., Goel, A. K., & Kumar, A. (2023). The role of artificial neural network and machine learning in utilizing spatial information. *Spatial Information Research*, *31*(3), 275–285. https://doi.org/10.1007/s41324-022-00494-x

Goh, G. D., Wong, K. K., Tan, N., Seet, H. L., & Nai, M. L. S. (2024). Large-format additive manufacturing of polymers: A review of fabrication processes, materials, and design. *Virtual and Physical Prototyping*, *19*(1), e2336160. https://doi.org/10.1080/17452759.2024.2336160

Gregoire, J. M., Zhou, L., & Haber, J. A. (2023). Combinatorial synthesis for AI-driven materials discovery. *Nature Synthesis*, *2*(6), 493–504. https://doi.org/10.1038/s44160-023-00251-4

Guo, K., Yang, Z., Yu, C. H., & Buehler, M. J. (2021). Artificial intelligence and machine learning in design of mechanical materials. *Materials Horizons*, *8*(4), 1153–1172. https://doi.org/10.1039/D0MH01451F

Habeeb, H. A., Wahab, D. A., Azman, A. H., & Alkahari, M. R. (2023). Design optimization method based on artificial intelligence (hybrid method) for repair and restoration using additive manufacturing technology. *Metals*, *13*(3), 490. https://doi.org/10.3390/met13030490

Huang, Z., Shen, Y., Li, J., Fey, M., & Brecher, C. (2021). A survey on AI-driven digital twins in industry 4.0: Smart manufacturing and advanced robotics. *Sensors*, *21*(19), 6340. https://doi.org/10.3390/s21196340

Janiesch, C., Zschech, P., & Heinrich, K. (2021). Machine learning and deep learning. *Electronic Markets*, *31*(3), 685–695. https://doi.org/10.1007/s12525-021-00475-2

Krauklis, A. E., Karl, C. W., Gagani, A. I., & Jørgensen, J. K. (2021). Composite material recycling technology—State-of-the-art and sustainable development for the 2020s. *Journal of Composites Science*, *5*(1), 28. https://doi.org/10.3390/jcs5010028

Liu, M., Li, H., Zhou, H., Zhang, H., & Huang, G. (2024). Development of machine learning methods for mechanical problems associated with fibre composite materials: A review. *Composites Communications*, 101988. https://doi.org/10.1016/j.coco.2024.101988

Marcon, É., Soliman, M., Gerstlberger, W., & Frank, A. G. (2022). Sociotechnical factors and Industry 4.0: An integrative perspective for the adoption of smart manufacturing technologies. *Journal of Manufacturing Technology Management*, *33*(2), 259–286. https://doi.org/10.1108/JMTM-01-2021-0017

Mazhar, T., Malik, M. A., Haq, I., Rozeela, I., Ullah, I., Khan, M. A., ... & Hamam, H. (2022). The role of ML, AI and 5G technology in smart energy and smart building management. *Electronics*, *11*(23), 3960. https://doi.org/10.3390/electronics11233960

Nath, D., Misra, M., Al-Daoud, F., & Mohanty, A. K. (2025). Studies on poly (butylene succinate) and poly (butylene succinate-co-adipate)-based biodegradable plastics for sustainable flexible packaging and agricultural applications: A comprehensive review. *RSC Sustainability*, *3*, 1267–1302. https://doi.org/10.1039/D4SU00193A

Negut, I., & Bita, B. (2023). Exploring the potential of artificial intelligence for hydrogel development—a short review. *Gels*, *9*(11), 845. https://doi.org/10.3390/gels9110845

Ng, W. L., Goh, G. L., Goh, G. D., Ten, J. S. J., & Yeong, W. Y. (2024). Progress and opportunities for machine learning in materials and processes of additive manufacturing. *Advanced Materials*, *36*(34), 2310006. https://doi.org/10.1002/adma.202310006

Preethikaharshini, J., Naresh, K., Rajeshkumar, G., Arumugaprabu, V., Khan, M. A., & Khan, K. A. (2022). Review of advanced techniques for manufacturing biocomposites: Non-destructive evaluation and artificial intelligence-assisted modeling. *Journal of Materials Science*, *57*(34), 16091–16146. https://doi.org/10.1007/s10853-022-07558-1

Qiu, C., & Yang, J. (2022). Machine learning applications in composites: Manufacturing, design, and characterization. In *Machine Learning in Materials Informatics: Methods and Applications*, pp. 65–85. ACS. DOI: 10.1021/bk-2022-1416.ch004

Raabe, D. (2023). The materials science behind sustainable metals and alloys. *Chemical Reviews*, *123*(5), 2436–2608. https://doi.org/10.1021/acs.chemrev.2c00799?urlappend=%3Fref%3DPDF&jav=VoR&rel=cite-as

Radu, I. C., Vadureanu, A. M., Cozorici, D. E., Blanzeanu, E., & Zaharia, C. (2025). Advancing sustainability in modern polymer processing: Strategies for waste resource recovery and circular economy integration. *Polymers*, *17*(4), 522. https://doi.org/10.3390/polym17040522

Sadat, T. (2023). Machine learning-assisted tensile modulus prediction for flax fiber/shape memory epoxy hygromorph composites. *Applied Mechanics*, *4*(2), 752–762. https://doi.org/10.3390/applmech4020038

Seckin, M., Demircioglu, P., Seckin, A. C., Bogrekci, I., & Aksoy, S. (2025). Artificial intelligence for non-destructive imaging in composite materials. *Eng*, *6*(3), 46. https://doi.org/10.3390/eng6030046

Sengottaiyan, S., Subbarayan, S., Viswanathan, V., & Pugazhendi, P. (2024). Optimized machine learning with hyperparameter tuning and response surface methodology for predicting tribological performance in bio-composite materials. *Polymer Composites*, *45*(10), 9421–9439. https://doi.org/10.1002/pc.28418

Shokrollahi, Y., Nikahd, M. M., Gholami, K., & Azamirad, G. (2023). Deep learning techniques for predicting stress fields in composite materials: A superior alternative to finite element analysis. *Journal of Composites Science*, *7*(8), 311. https://doi.org/10.3390/jcs7080311

Simões, S. (2024). High-performance advanced composites in multifunctional material design: State of the art, challenges, and future directions. *Materials*, *17*(23), 5997. https://doi.org/10.3390/ma17235997

Skosana, S. J., Khoathane, C., & Malwela, T. (2025). Driving towards sustainability: A review of natural fiber reinforced polymer composites for eco-friendly automotive light-weighting. *Journal of Thermoplastic Composite Materials*, *38*(2), 754–780. https://doi.org/10.1177/08927057241254324

Tran, H., Gurnani, R., Kim, C., Pilania, G., Kwon, H. K., Lively, R. P., & Ramprasad, R. (2024). Design of functional and sustainable polymers assisted by artificial intelligence. *Nature Reviews Materials*, 1–21. https://doi.org/10.1038/s41578-024-00708-8

Uddin, M. H., Mulla, M. H., Abedin, T., Manap, A., Yap, B. K., Rajamony, R. K., ... & Nur-E-Alam, M. (2025). Advances in natural fiber polymer and PLA composites through artificial intelligence and machine learning integration. *Journal of Polymer Research*, *32*(3), 76. https://doi.org/10.1007/s10965-025-04282-7

Umeliwu, O. A., & Samuel, B. O. (2024). Challenges and Opportunities of Artificial Intelligence on composite material engineering. *Journal of Sustainable Materials Processing and Management*, *4*(2), 61–76. Available at: https://penerbit.uthm.edu.my/ojs/index.php/jsmpm/article/view/18750

Wang, J., Li, Y., Gao, R. X., & Zhang, F. (2022). Hybrid physics-based and data-driven models for smart manufacturing: Modelling, simulation, and explainability. *Journal of Manufacturing Systems*, *63*, 381–391. https://doi.org/10.1016/j.jmsy.2022.04.004

Westermayr, J., Gastegger, M., Schütt, K. T., & Maurer, R. J. (2021). Perspective on integrating machine learning into computational chemistry and materials science. *The Journal of Chemical Physics*, *154*(23). https://doi.org/10.1063/5.0047760

Yakoubi, S. (2024). Sustainable revolution: AI-driven enhancements for composite polymer processing and optimization in intelligent food packaging. *Food and Bioprocess Technology*, 1–26. https://doi.org/10.1007/s11947-024-03449-2

Yao, Z., Lum, Y., Johnston, A., Mejia-Mendoza, L. M., Zhou, X., Wen, Y., ... & Seh, Z. W. (2023). Machine learning for a sustainable energy future. *Nature Reviews Materials*, *8*(3), 202–215. https://doi.org/10.1038/s41578-022-00490-5

7

IoT-Based Algorithms for Robots: A Case Study on Remanufacturing

Anupam Bonkra, Jai Sonker, Harvinder Singh, Rahul Grover, Ankit Sharma, Sriharsha Chollangi, and Padmini Yenumula

7.1 Introduction

The dawn of the Fourth Industrial Revolution, also known as Industry 4.0, has reshaped the industrial landscape through the integration of advanced digital technologies (Kumar et al., 2019). Among the most transformative of these are robotics, artificial intelligence (AI) (Bonkra et al., 2023), and the Internet of Things (IoT) (Bonkra & Dhiman, 2021). Together, these technologies are driving the shift from traditional manufacturing models toward intelligent, data-driven, and highly automated production environments. One of the most important areas where the changes of this transformation will find serious opportunities for taking advantage is remanufacturing. In fact, remanufacturing is the process of returning used, worn-out, or discarded products to a "like-new" condition typically defined to be no worse than with respect to the original specifications while often actually achieving performance equal to or better than the original specifications. Remanufacturing thus avoids the breakdown of products into raw materials, but entails keeping alive value embedded within existing parts, embarking upon the following: cost savings, lower environmental footprint, and long product life. The very nature of remanufacturing, however, creates some formidable hurdles. The varying degrees of wear of incoming products, unpredictable defects, and customized disassembly and reassembly are huge challenges to automation and efficiency of operations for this extent. Here, this is a convergence of IoT and robotics, which brings a game-changing advantage. The IoT-based algorithm can be embedded in robotic systems to let machines communicate in real time with their environment, learn from data, and adapt to unexpected changes. IoT connects robots with sensors, actuators, and cloud connectivity and lets them collect and analyze millions of terabytes of data concerning the products they handle. All these make real-time decision-making, context-aware task execution, and even predictive maintenance

DOI: 10.1201/9781003506492-7

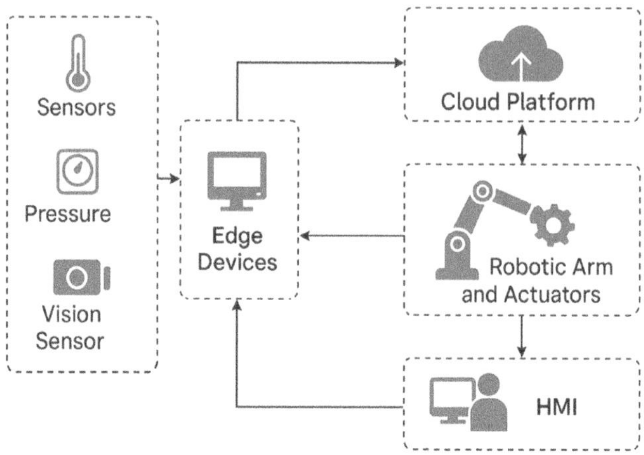

FIGURE 7.1
IoT-enabled robotic remanufacturing system architecture.

possible to ensure that robots themselves operate on an optimum level. Furthermore, IoT also allows collaborative robotics, or "cobot," to work efficiently and safely with human operators. These smart systems are capable of evaluating a part, determining the best disassembly method, selecting the appropriate tools, and refurbishing or reassembling components with minimal human intervention. Thus, remanufacturing will have a faster, better, and greener process. This chapter talks about the transformative role of IoT in robotic remanufacturing. The chapter contains a real-world case study to show how these algorithms have been implemented within a robotic system to optimize several different stages of the remanufacturing workflow. Readers will gain an in-depth understanding of the architecture, benefits, and challenges concerning the deployment of IoT-enabled robotics in a highly complex industrial environment by looking at this case. The architecture of IoT-enabled robotic remanufacturing integrates sensor data, edge computing, and cloud platforms to enhance automation and decision-making in remanufacturing systems (Figure 7.1).

7.2 Understanding IOT in Robotics

7.2.1 What Is IoT?

The IoT (Tiwana & Singh, 2023) is a rapidly expanding network that connects thousands of physical objects or actual items—from sensors and actuators to everyday machines—embedded with software, communication technologies, and data analytics capabilities, and builds real-time data collection, exchange, and interpretation devices. All these real-time data can be used for informed decisions-making by the IoT systems, change detections, and action executions with little or without human intervention. Automation of machines is increased while IOT-integrated robotics simulates them as part of a big entire digital ecosystem which uses new procedures, rather than isolated units, and operates as intelligent agents. Then robotic systems can interpret conditions existing in the real world and adapt their behaviors according to their tasks with higher precision and autonomy with much source data being such as temperature sensor data, camera data, GPS modules data, and cloud platform inputs. For instance, in a remanufacturing environment, an IoT-enabled robot would detect an out-of-rules condition involving a component, refer to cloud-based diagnostic data, and then initiate appropriate disassembly or repair action determined by the current actual information. Such synergistic benefits will cut inefficiencies associated with conventional operations, while enhancing advanced capabilities such as remote monitoring, real-time diagnostics, and predictive maintenance in the case of IoT-enabled robotic capabilities. Therefore, IoT robotic systems now become necessary in a highly complex environment where adaptability intelligence and interconnectivity are fundamental for success.

7.2.2 Role of IoT in Industrial Robotics

The IoT is an extraordinary leap into the future of modern manufacturing and automation, with integration into industrial robotics. Industrial robots have usually existed in seclusion, assigned with a set of operations and starkly limited adaptability along the monotony of their program. However, with the advent of IoT technologies, these robots are now smart, connected systems that can rely on their external environment, do data-driven decisions, and interact not only with machines but also with humans. Real-time information collection and communication are perhaps some most important roles of IoT for industrial robotics. Robots, by using RFID tags (Wang et al., 2021), vision systems, and various other IoT-enabled devices, can put up a continuous monitoring of parameters like temperature, vibration, pressure, and wear on components. Data then are transferred on communications network to centralized systems, edge devices, or cloud platforms for further analysis and decision-making. This real-time data would allow immediately reacting to changes in the environment, increasing accuracy and reliability during operation. Adaptive process control is another significant contribution made by IoT. In most such dynamic industrial environments, such as remanufacturing, the products are subjected to lots of variability of changes and many unexpected events occur. Under such conditions, it is IoT's importance for robots to change the way they are doing things on-the-fly depending on the particular state or configuration of a component. For instance, modifications can be made on the disassembly process even if the part is found to be misaligned or damaged—this truly increases flexibility levels and in turn reduces downtime. More than that, predictive maintenance is now possible with IoT. By constant monitoring of the healthy status of all robotic components, including motors, joints, and end-effectors, IoT systems will help to forecast future failures and determine when machinery will break down. Thus, sudden breakdowns can be avoided; maintenance costs can be reduced and equipment can last longer. Maintenance teams can be threatened with advanced alerts and performance reports through dashboards that enable a more proactive and data-centered approach to maintenance. Furthermore, remote access and control are enabled by IoT in robotics. Now, operators and engineers can monitor robot performance metrics, do diagnostics, and even perform some functions from remote locations through secure interfaces. It makes sense in dangerous or otherwise hard-to-reach places where access is not only risky but also sometimes impossible. Finally, IoT paves the way for working together, using robots without human operators, and many industrial systems work in the background. Direct and indirect collaboration between robots and other systems will take place over an Internet connection by connecting robots with enterprise resource planning systems, manufacturing execution systems, and supply chain management platforms. Thus, horizontal coordination can be achieved along the entire production line. Robots will be able to receive task updates and shift schedules as required with respect to demand and can report process progress in real time toward supporting smarter and agile production workflows.

7.2.3 Real-Time Communication between Robotic Arms, Sensors, and Control Systems

Real-time communication is a key element in assuring the efficient operation of IoT-based industrial robots. This is concerned with unending and instantaneous data interaction between any kinds of robotic components—robotic arms, embedded sensors acting as sentinels for the state of their environment, and centralized or distributed control systems. This communication architecture ensures synchronization, responsiveness, and the ability to perform a given task under extreme precision with minimum delay by all parts of the robotic system. In any standard industrial scenario, robotic arms are supported by several kinds of sensors such as force, temperature, vision, and proximity sensors that observe the environment with respect to the robot's internal condition. These sensors generate data continuously as they are operating. To ensure precise output from the robotic system, such sensor data has to be passed on to the control system in real time, where the data is processed and commands are sent back to the robotic arms accordingly. This is where the role of IoT comes in, providing a basis for the infrastructure for the reliable and low-latency transfer of information. IoT smooths out such data exchanges among devices using industrial Ethernet, wireless sensor networks, and edge computing devices without significant lag. In this way, the real-time data loop enables robots to adjust their movements to avoid collisions with speed, reacting to changes in an object's position or orientation, and to coordinate with other machines in the production line. Moreover, the real-time charm with communication empowers intelligence in robotic systems on a distributed nature. Rather than depend on the central controller alone to make decisions for the components, each will decide on its own, thereby speeding up the operations through local data. For example, in a remanufacturing environment, if one robotic arm detects a problem during disassembly, it can alert all other components in the system to either halt or change their action, thereby averting harm or error.

7.2.4 Data-Driven Optimization for Operational Efficiency

Achieving operational efficiency in industrial robotics is highly prioritized. Using the IoT technologies, robotic systems can gather and analyze massive amounts of data, and can thus make informed decisions to enhance the productivity, accuracy, and resource utilization of the production processes. This is called data-driven optimization, where robots are enabled to go beyond pre-programmed routines and adaptive learning towards improved performance based on real-time and historical data. IoT-integrated robotic systems possess a whole array of sensors and communication modules, all pulling data from every corner of the production process—be it machine performance measurement, environmental conditions, cycle times, energy consumption, quality evaluation of products, etc. On the one hand, these data come together on analytical platforms or edge computing devices, where they are analyzed to recognize patterns, flag inefficiencies, and make suggested improvements. For example, in a remanufacturing setup, a robot could perform disassembly on components repeatedly; the condition and structure of those components may vary. This information—collected and analyzed—could allow the system to find out the most helpful methods and choose the one that shortens disassembly time and tool wear or enhances accuracy for better throughput with minimum downtime. Predictive analytics are another aspect of data-driven optimization: The system could trend over time and foresee any interruptions, such as declining performance trends of a robotic arm or increasing speed defect rate for a certain task. Such foreknowledge will prepare the system for relevant interventions to stem larger disruption and steady workflow. Also, aggregating data from several robots and systems could allow higher operations to be optimized. Task scheduling algorithms can allocate production resources more efficiently, balance workloads among machines, and minimize idle time based on real-time production data. Similarly, energy management systems can maximize the efficiency of power consumption during low demand periods.

7.2.5 Integration with Cloud and Edge Computing for Faster Decision-Making

The connectivity of the cloud with the edge of IoT-enabled robotics is an important aspect of faster decision-making in industrial environments. With respect to cloud computing, it should provide centralized data storage, high processing power, and advanced analytics and machine learning models that an intelligent robot would use to send operational data to the cloud for deep analysis. System-wide coordination and long-term optimization would be enabled by the cloud. But in cases requiring instantaneous responses, such as stopping a robotic arm to prevent a collision, any delay associated with a cloud-based decision can become a hindrance. This is where edge computing comes into the picture—a cloudless environment where the computing resources would exist closer to the data and the edge devices would quickly process data locally and make real-time decisions without reference to any external server. This combination will leverage both cloud and edge computing: for instance, while the cloud is doing complex, non-on-demand tasks, edge systems would ensure long continuity and low latency replies to critical needs. This means that you can balance the infrastructure in order to make industrial robotic systems more responsive, more intelligent, and more adaptable to situations. The IoT-enabled robotic remanufacturing process involves a systematic flow from core identification to quality control, enhancing efficiency and traceability (Figure 7.2).

7.3 Overview of Remanufacturing

Remanufacturing is an advanced industrial operation that restores used, worn-out, or end-of-life products to a condition that is at least equivalent to, or even better than, the original one in terms of performance and quality (Khan et al., 2025). It is a standardized and comprehensive industrial process that encompasses inspection, disassembly, cleaning, reconditioning, replacement of damaged parts, reassembly, and final testing. The goal is to give the product a warranty equal to that of a new product and its sale in "like new" condition. This is of utmost importance in industries like automotive, aerospace, heavy machinery, and electronics, where equipment is very costly and where sustainability is becoming an ethos. Remanufacturing benefits the circular economy by minimizing waste, using less raw material, and conserving energy. Remanufacturing thus reduces environmental impact and provides economic value by enhancing product lifecycle. Remanufacturing faces its challenges, however. Returning products may be in highly variable conditions, calling for flexible processes and skilled labor to handle different degrees of wear and damage. Often labor-intensive,

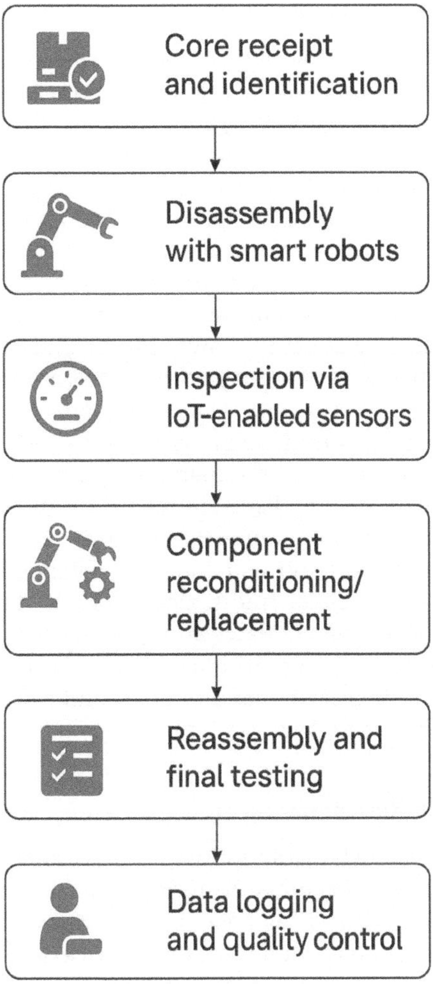

FIGURE 7.2
IoT-enabled robotic remanufacturing process flow.

remanufacturing processes are less automated compared with modern product manufacture, making an integration of advanced technologies like robotics, AI, and the IoT into remanufacturing operations more attractive. Such technologies promise to bring precision, efficiency, and reliability into the process, which in turn can mitigate unpredictability and complexity that have been long limiting the scalability of remanufacturing. As the industry moves toward sustainability and cost-effective means of production, remanufacturing will have a central role to play, buoyed by innovations that facilitate smarter, more adaptive, and environmentally conscious operations.

7.3.1 Importance of Remanufacturing

Remanufacturing is one of the bigger circumscribers of worlds on the efficiency side and environmentally acceptable aspects of new industries. The future of remanufacturing seems brighter as demand for sustainable practices increases across the world. Remanufacturing offers practical solutions that allow increased light on the life cycle of an asset by lowering the need for raw materials and waste generation. The preservation of natural resources would therefore provide much more energy at a lesser cost and less carbon emission than separately manufacturing new products from scratch. Economically, this provides firms a chance of value recovery from returned goods or end-of-life products, which allows them to lower production costs while providing their consumers with goods of high quality at a lower price. - Industries, particularly those reliant on critical or scarce components, also gain expanded opportunities to manage supply chain risks and shortages effectively. This includes localized job creation as remanufacturing requires highly trained workers to perform inspection, disassembly, reconditioning, and quality control. Remanufacturing, in the more extensive concept, thus fosters a circular economy—a model rooted in reuse, repair, and regeneration—to help industries turn away from the linear traditional model of "take, make, dispose." As the tightening environmental laws and growing consumer awareness set in, remanufacturing is fast gaining importance as a means through which businesses can fulfill their sustainability goals and remain competitive in a world constrained by resources.

7.3.2 Challenges in Remanufacturing

Remanufacturing is a sustainable industrial procedure that includes disassembling used products—or cores—restoring their components, and reassembling them to create products "as good as new" or even better. This process plays a vital role in circular economy strategies as it extends the productive life of a product and also reduces environmental impact and conserves resources. However, despite its advantages, remanufacturing is accompanied by several challenges and bottlenecks that slow its efficient take-up and implementation. These challenges range from technical and economic to regulatory and environmental, and each must be addressed for increasing the feasibility and scalability of remanufacturing operations.

7.3.2.1 Product Design Not Suited for Remanufacturing

With remanufacturing, one of the major problems is that most products are not manufactured for remanufacturing in the first place. Manufacturers almost always put more weight on price, performance, and appearance than on ease of disassembly or reuse of components. This results in complex assemblies with non-modular design, strong adhesive, or irreversible joining like welding, making it difficult to take them apart without inflicting damage. A redesign for remanufacturing, using modular components with standardized interfaces, would greatly advance disassemblability and component recovery.

7.3.2.2 Uncertainty in Core Quality and Supply

Remanufacturing operations heavily depend on the availability and condition of returned products, known as cores. However, there is often significant uncertainty regarding when, how many, and in what condition these cores will be returned. The quality of cores can vary widely due to differences in usage, maintenance, and wear. This unpredictability makes it difficult to plan production schedules, manage inventory, and ensure consistent output quality. Implementing better tracking and forecasting systems for core returns can help address this issue.

7.3.2.3 High Disassembly and Inspection Costs

The entire process of disassembly and inspection is the most labor-intensive in remanufacturing. Disassembling products from use and inspecting their individual components for damage or wear is, however, done mostly by humans and not on automated production lines. This process could prove very costly and time-consuming, particularly using advanced non-destructive testing techniques to assess the reliability of the examined parts. But over time, costs will decrease when investing in automation technologies such as robotics and AI-driven inspection tools.

7.3.2.4 Lack of Standardized Processes

The remanufacturing industry also suffers from the absence of standard methods and practices for ensuring quality. Due to variations in cores and a shortage of guidelines recognized by the industry, remanufacturing processes may vary considerably from facility to facility, even for the same category of product. This lack of standardization is a major contributor to quality variations causing mistrust among customers. If globally accepted standards and certifications for remanufactured products were to be established, the issue of quality control would receive a big boost, and in turn, increase market acceptance.

7.3.2.5 Technological Obsolescence

With most return products now out of date or not compatible with current systems and regulations, in certain instances the components become obsolete, which creates a hindrance in sourcing replacements and complying with the new standards. In some cases, remanufacturers must either upgrade or retrofit the products using new technology, which can become complex and costly. Monitoring industry trends and flexible manufacturing processes can be of relevance in this regard.

7.3.2.6 Economic Viability and Market Perception

Economic viability of remanufacturing hinges mainly on whether a cost-efficient system can be developed for resale. Labor, inspection, and reassembly could be so high that a remanufactured item is seldom less expensive than that which is new. Further, there is often still an issue of public perception—many think of remanufacturing as inferior or as an uncertain product. This stigma can limit demand and affect pricing. So in a way, it could even be a marketing issue, and moreover, educating consumers as well as offering a warranty or certification would be very valuable in building the trust factor for remanufactured products.

7.3.2.7 Regulatory and Legal Issues

Various challenges in law and regulations are imposed on remanufacturing. These include, among others, electronics, healthcare, and automotive sectors. Environmental regulations on the disposal of materials, emissions, and hazardous substances are included in addition to the aspects of intellectual property relating to certain proprietary designs or software. Crossing these complex legal landscapes demands attention to compliance, documentation, and possibly expensive legal support for cross-border remanufacturing.

7.3.2.8 Environmental Impact and Waste

Remanufacturing is normally an environment-friendly aspect, but not all materials or components can be reused. There will always be residual waste with unusable parts which should be managed responsibly. Improper handling entails environmental contamination, especially with hazardous substances. Cleaning, repairing, and testing consume energy that all counts against the environment. Adopting better practices and developing recycling schemes for unrecyclable components can help mitigate damage.

7.3.2.9 Skilled Labor Shortage

Remanufacturing calls for a labor force prepared in the specific tasks of diagnostics, disassembly, testing, and reassembly. Oftentimes, there is a shortage of personnel in these fields, particularly with automation encroaching upon traditional manufacturing. Training programs and consortiums with institutes of learning can develop a pipeline for skilled labor, assuring that the industry has human resources to meet ever-increasing demand.

7.3.2.10 Logistics and Reverse Supply Chain Management

Efficient remanufacturing requires an efficient reverse supply chain to collect, transport, and store returned products. The management of reverse flow is, however, inherently more complex than that of traditional supply chains, possessing varying degrees of uncertainty with respect to time, condition, and volume. Also, reverse logistics can be expensive, particularly for bulky or heavy items. Thus, deploying tracking systems and optimizing routing would provide a better mixed benefit of efficiency and cost advantage to reverse logistics activities.

7.4 Case Study: IoT-Enabled Robotic System in Remanufacturing

The global deployment of IoT-enabled robotic remanufacturing is expanding across aerospace and automotive sectors, as illustrated in the country-wise map (Figure 7.3). Table 7.1 presents a comprehensive overview of country-specific implementations of IoT-enabled robotic remanufacturing systems, illustrating diverse technological approaches and associated benefits across industries.

7.5 Benefits and Outcomes

IoT-enabled robotic systems incorporated into remanufacturing processes have indeed transformed and benefited the entire globe. This made an increase in efficiency through all kinds of work done in different companies and countries. Automated work such as disassembly, inspection, and reassembly that ordinarily required manual labor has now been taken over by these wonderful systems. When IoT is deployed in real-time data, it produces a condition monitoring and predictive maintenance environment. This enables the prediction of unplanned downtimes, enhances the reliability of remanufactured components, and allows robots and IoT systems to support their respective analytics as part of operational features. Additionally, smart remanufacturing supports the goals of establishing sustainability by maximizing resource recovery and minimizing wastes at the same time while extending the lifespan of critical assets. It optimizes resource utilization, shortens lead times, and enables improved traceability to the companies with digital twin and data-driven decision-making models. Not only are these improvements beneficial cost- and productivity-wise, but they will

FIGURE 7.3
Country-wide deployment map of IoT-enabled robotic remanufacturing in aerospace and automotive sectors.

also create new ways whereby businesses can stay competitive in the face of ever-tightening environmental rules and consumer expectations by promoting circular economy practices. Several emerging technologies, including blockchain, edge AI, multi-robot coordination, and protocol standardization, are driving the evolution of IoT-enabled robotic remanufacturing (Figure 7.4).

7.6 Challenges and Limitations

The IoT notions have been ameliorated at a large extent; there are still several challenges and limitations regarding their implementation in the remanufacturing sector and exact operations. High initial investment costs for robotics and IoT infrastructure count as a main barrier against small and medium enterprises, as these in many cases cannot absorb the investment cost. Interoperability issues between legacy systems and modern IoT platforms create further obstacles to smooth integration and data flows across the operations. The other pressing issue is perhaps in terms of cybersecurity; interconnected systems can be .vulnerable to all sorts of breaches and attacks that can lead to disruption in operations. Besides, remanufacturing involves working with used and variable products, which adjust current robotic capabilities; this is in turn an active area of growth. The collection of data management and analytics comes next; generating huge amounts of real-time data will help only if processed and utilized timely, which in itself requires

TABLE 7.1

Country-Wise IoT-Enabled Robotic Remanufacturing Case Studies

Country	Organization/ Case	IoT Technologies Used	Type of Robots	Remanufactured Products	Application/Process Description	Key Benefits
USA (Poon, 2021)	Caterpillar Reman	Embedded sensors, real-time analytics	Industrial robotic arms	Engine blocks, hydraulic pumps	Automated disassembly, condition-based sorting, and cleaning	Increased throughput, reduced inspection errors
Germany (Artiushenko et al., 2024)	Bosch Rexroth	Edge IoT devices, predictive maintenance platforms	Multi-axis robotic arms	Hydraulic valves and cylinders	Robotic cells disassemble and test hydraulic components	Optimized energy use, reduced downtime
Japan (Zarrar et al., 2021)	Toyota Industries	IoT-enabled MES, sensor networks	Collaborative robots (cobots)	Car engines, electric motors	Cobots assist in reassembly; IoT tracks part conditions and usage	Process consistency, improved worker safety
China (Wu et al., 2024)	Haier Smart Factory	RFID, IoT-based MES, cloud computing	Robotic arms, AGVs	Washing machines, refrigerators	Smart robots sort, inspect, and assemble recovered components	Reduced cycle time, better resource management
South Korea (Muralidharan et al., 2016)	LG Electronics	IoT telemetry, sensor arrays, cloud analytics	Robotic arms, automated testers	Monitors, televisions	IoT guides part diagnostics; robotics handle modular repairs	Data-driven maintenance, faster turnaround
India (Packianathan et al., 2025)	Tata Motors	IoT dashboards, condition monitoring	Pick-and-place robots, cleaning robots	Vehicle transmissions and engines	Automated disassembly and component cleaning guided by sensor data	Reduced manual labor, better fault detection
UK (Hegazy Abdelghany Mohamed Ammar, 2021)	Rolls-Royce	Digital twin, IoT sensors, edge computing	Robotic arms, inspection drones	Jet engine blades and components	Inspection drones feed IoT systems; robots handle precision tasks	Enhanced quality control, lifecycle visibility
France (Gokhale & Iyer, 2025)	Renault Re-Factory	RFID tracking, cloud-based analytics	Cobots, AGVs	EV batteries, powertrains	IoT tracks part reuse, robotics disassemble and recondition systems	Extended product life, improved traceability

(continued)

TABLE 7.1 (Continued)
Country-Wise IoT-Enabled Robotic Remanufacturing Case Studies

Country	Organization/ Case	IoT Technologies Used	Type of Robots	Remanufactured Products	Application/Process Description	Key Benefits
Italy (Ionescu et al., 2025)	Comau (FIAT Group)	IoT sensors, PLC integration	Automated robotic cells	Automotive engines and gearboxes	Robots disassemble and refurbish with real-time feedback	Higher productivity, real-time fault detection
Canada (Eskue, 2023)	Bombardier Aerospace	Structural health monitoring (IoT), digital thread	Precision robotic tools	Aircraft landing gear, actuators	IoT evaluates structural integrity; robots assist in remanufacturing	Predictive part replacement, high precision
Brazil (Reis et al., 2019)	Embraer	Sensor networks, IoT predictive systems	Robotic arms, precision testers	Hydraulic and mechanical components	IoT sensors guide robotic repair of aviation parts	Increased reliability, compliance with aviation standards
Australia (Keung et al., 2022)	CSIRO Smart Factory	Cloud-based IoT, edge AI sensors	Mobile robotic systems	Mining equipment parts	Smart inspection and remanufacturing using adaptive robotics	Reduced material waste, faster diagnostics
Sweden (Larsson & Bengtsson, 2022)	Volvo Group	IoT lifecycle analytics, smart sensors	Collaborative robots	Truck axles, gearboxes	IoT tracks product use; cobots assist in modular remanufacturing	Lower energy use, reduced human error
Netherlands (Turskis & Šniokienė, 2024)	Philips Circular Factory	IoT-driven traceability systems	Assembly/ disassembly robots	Medical imaging and diagnostic equipment	Smart tracking of parts, automated reassembly of certified components	Regulatory compliance, patient safety
Singapore (Zafar et al., 2024)	ARTC	Real-time IoT monitoring, digital platforms	Dual-arm robots, cobots	Aerospace parts, electronics	Robotics integrated with IoT for real-time status and process feedback	Shorter lead times, data-based decisions
South Africa (Lorimer et al., 2019)	Denel Vehicle Systems	Embedded diagnostics, remote IoT monitoring	Robotic manipulators	Armored vehicle components	Robots perform part evaluation and restoration under IoT control	Enhanced fleet readiness, safer operations
UAE (Ali & Anjum, 2025)	Dubai Smart Industry	Smart factory IoT platforms, AI analytics	Autonomous robots, robotic arms	Industrial motors, turbines	IoT data guides robotic repair operations and failure prediction	Energy efficiency, proactive maintenance

Russia (Zecha et al., 2013)	United Engine Corp.	IoT telemetry, vibration and heat sensors	High-precision robotic arms	Turbine engine blades, rotors	Robotic reconditioning informed by real-time IoT diagnostics	Improved performance, reduced rework
Mexico (Garrido-Hidalgo et al., 2023)	Siemens Remanufacturing	Smart sensors, digital IoT twin	Robotic assembly cells	Industrial drive systems	IoT ensures exact reassembly; robots standardize quality	Higher reliability, predictive insights
Malaysia (Ibrahim, 2017)	Proton Smart Factory	IoT remote diagnostics, smart inventory	Collaborative robots	Hybrid car batteries, ECUs	IoT identifies reusable parts, cobots manage safe battery handling	Cost savings, better worker ergonomics

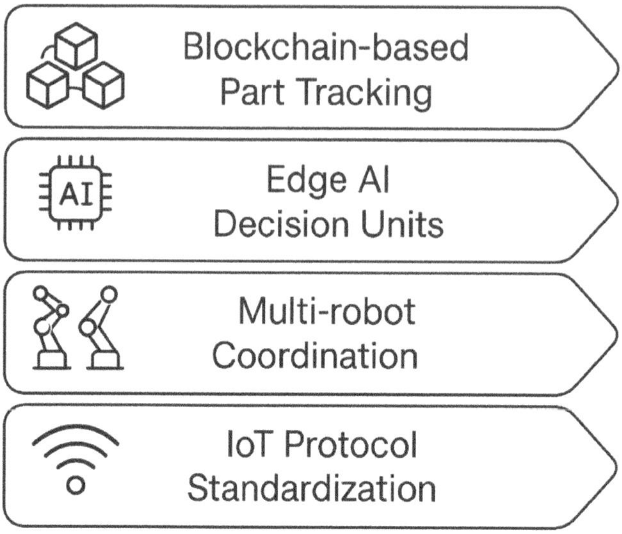

FIGURE 7.4
Key technological enablers for advanced IoT-enabled robotic remanufacturing systems.

skilled personnel and advanced AI models. And then, if we talk about skill gaps, many of them do not even have proper training to operate and maintain smart robotic systems. Finally, industry-specific regulations and compliance issues, especially in sensitive industries like healthcare and aviation, create barriers in the adoption of standardized IoT–robotic solutions. These constraints show the need for strategic planning, workforce upskilling, and resilient digital infrastructure in order to successfully seize the opportunities offered by IoT-enabled robotic remanufacturing.

7.7 Future Directions

Future opportunities in research and development for industry IoT-enabled robots remanufacturing standards would focus on some areas that have shown promise in overcoming existing constraints and enhancing system capabilities. One such thrust is the adoption of the blockchain technology for its role in providing a secure and transparent tracking regime on remanufactured parts across the entire lifecycle. This would immensely improve traceability and thereby trust in manufactured products in secondary markets. Another area is the edge AI for faster, localized decision-making by processing data at a point closer to source, thereby reducing latency and dependence on cloud infrastructure. Multi-robot collaboration advances could increase the handling capacity of coordinated robotic systems for larger and more complex assemblies, thus promoting flexibility and efficiency in remanufacturing lines. Last but not least, a standard set of IoT protocols for various remanufacturing domains would be a prerequisite for ensuring compatibility and seamless data exchange between smart systems while at the same time assuring scaled capacity of these systems. These innovations would usher in an intelligent, secure, and, above all, sustainable future remanufacturing ecosystem.

7.8 Conclusion

Automation through IoT network framework is now changing the industrial way of remanufacturing by making processes move from traditional laborious and uncertain tasks to highly data-driven automated and intelligent workflows. Such real-time data collection, adaptive control, as well as predictive maintenance not only increase process efficiency within and among IoTs but also enable product quality and environmental sustainability. Real-world case examples of the research explore how different countries and sectors are using this industrial technology to confront remanufacturing limitations that have long existed and then accessing new points of performance. Certainly, while durability and the efficiency of IoT and robotics bring with them their own burdens like data security risks, high setup costs, or standardization needs, recent exciting advancements into standards, blockchain technologies, edge computing, and collaborative robotics still hold under application for future synergetic solutions. Also, for industrial beginning with deeper digital transformation, remanufacturing will play a vital role in an economical circle to meet long-term economic and ecological goals.

References

Ali, H. S. W., & Anjum, A. (2025). A comprehensive review regarding Industry 4.0, IoNT and IoT for smart manufacturing betterment. *Kashf Journal of Multidisciplinary Research*, 2(01), 46–56.

Artiushenko, V., Lang, S., Lerez, C., Reggelin, T., & Hackert-Oschätzchen, M. (2024). Resource-efficient Edge AI solution for predictive maintenance. *Procedia Computer Science*, 232, 348–357.

Bonkra, A., & Dhiman, P. (2021, December). IoT security challenges in cloud environment. In *2021 2nd international conference on computational methods in science & technology (ICCMST)* (pp. 30–34). IEEE.

Bonkra, A., Bhatt, P. K., Rosak-Szyrocka, J., Muduli, K., Pilař, L., Kaur, A., … & Rana, A. K. (2023). Apple leave disease detection using collaborative ML/DL and artificial intelligence methods: Scientometric analysis. *International Journal of Environmental Research and Public Health, 20*(4), 3222.

Eskue, N. (2023). Digital thread roadmap for manufacturing and health monitoring the life cycle of composite aerospace components. *Aerospace, 10*(2), 146.

Garrido-Hidalgo, C., Roda-Sanchez, L., Javier Ramírez, F., & Olivares, T. (2023, October). IoT-driven digital twin for improved product disassembly in remanufacturing. In *International Workshop on Autonomous Remanufacturing* (pp. 281–291). Cham: Springer Nature Switzerland.

Gokhale, N., & Lyer, C. V. H. (2025). Sustainability in the Digital Age: How marketing strategies and financial innovation drive circular technologies. In *Innovating Sustainability Through Digital Circular Economy* (pp. 305–328). IGI Global Scientific Publishing.

Hegazy Abdelghany Mohamed Ammar, M. (2021). *Advanced digital twins for conditions monitoring, examinations, diagnosis and predictive remaining lifecycles based Artificial Intelligence* (Doctoral dissertation, Brunel University London).

Ibrahim, M. H. (2017). *Smart manufacturing: Its challenge and future implementation in Malaysia* (Master's thesis, University of Malaya (Malaysia)).

Ionescu, D., Filipescu, A., Simion, G., & Filipescu, A. (2025). IoT-Cloud and digital twin approach for HIL structure based on inverse kinematics model and PLC for controlling a robotic cell in the framework of Industry 4.0. MDPI.

Keung, K. L., Chan, Y. Y., Ng, K. K., Mak, S. L., Li, C. H., Qin, Y., & Yu, C. W. (2022). Edge intelligence and agnostic robotic paradigm in resource synchronisation and sharing in flexible robotic and facility control system. *Advanced Engineering Informatics, 52*, 101530.

Khan, W. A., Esat, V., Hammad, M., Ali, H., Zafar, M. Q., & Ali, R. (2025). Remanufacturing. In *Computer Aided Engineering Design and Manufacturing: A Fourth Industrial Revolution Perspective* (pp. 187–211). Springer Nature Switzerland.

Kumar, K., Zindani, D., & Davim, J. P. (2019). *Industry 4.0: Developments towards the fourth industrial revolution*. Springer.

Larsson, S., & Bengtsson, K. (2022). Enabling human-robot collaboration and intelligent automation in the automotive industry: A study of stakeholders perspectives.

Lorimer, T., Mouton, H. D., Smit, W. J., Van Niekerk, T., Luwes, N., Stopforth, R., … & Kruger, K. (2019). A review of robotics research in South Africa. *R&D Journal, 35*, 75–97

Muralidharan, S., Roy, A., & Saxena, N. (2016). An exhaustive review on Internet of things from Korea's perspective. *Wireless Personal Communications, 90*, 1463–1486.

Packianathan, R., Arumugam, G., Malaiarasan, A., & Natarajan, S. K. (2025). Integrating industrial robotics and internet of things (IoT) in smart transportation system. In *Driving Green Transportation System Through Artificial Intelligence and Automation: Approaches, Technologies and Applications* (pp. 379–395). Springer Nature Switzerland.

Poon, R. J. M. (2021). *Design and Control of a Mounted Robotic Arm Tool Changer and Measurement Tools for Agriculture* (Doctoral dissertation, Massachusetts Institute of Technology).

Reis, R., Diniz, F., Mizioka, L., Yamasaki, R., Lemos, G., Quintiães, M., … & Pereira, A. (2019). FASTEN: EU-Brazil cooperation in IoT for manufacturing. The Embraer use. In *MATEC Web of Conferences* (vol. 304, p. 04007). EDP Sciences.

Tiwana, P. S., & Singh, J. (2023, January). Load optimization accessions, ramification the QoS in software defined networking. In *2023 International Conference on Artificial Intelligence and Smart Communication (AISC)* (pp. 1013–1017). IEEE.

Turskis, Z., & Šniokienė, V. (2024). IoT-driven transformation of circular economy efficiency: An overview. *Mathematical and Computational Applications*, *29*(4), 49.

Wang, H., Yu, R., Pan, R., Liu, M., Huang, Q., & Yang, J. (2021). Fast tag identification for mobile RFID robots in manufacturing environments. *Assembly Automation*, *41*(3), 292–301.

Wu, K., Xu, J., & Zheng, M. (2024). Industry 4.0: Review and proposal for implementing a smart factory. *The International Journal of Advanced Manufacturing Technology*, *133*(3), 1331–1347.

Zafar, M. H., Langås, E. F., & Sanfilippo, F. (2024). Exploring the synergies between collaborative robotics, digital twins, augmentation, and industry 5.0 for smart manufacturing: A state-of-the-art review. *Robotics and Computer-Integrated Manufacturing*, *89*, 102769.

Zarrar, A., Rasool, M. H., Raza, S. M. M., & Rasheed, A. (2021, September). IoT-enabled lean manufacturing: Use of IoT as a support tool for lean manufacturing. In *2021 International Conference on Artificial Intelligence of Things (ICAIOT)* (pp. 15–20). IEEE.

Zecha, C. W., Link, J., & Claupein, W. (2013). Mobile sensor platforms: Categorisation and research applications in precision farming. *Journal of Sensors and Sensor Systems*, *2*(1), 51–72.

8

AI-Based Scientometric Analysis of Post-Processing in Additively Manufactured Structure: Insights from 2013 to 2023

Hreetabh Kishore, Priyanka Sinha, Muralimohan Cheepu, and Suresh Pratap

8.1 Introduction

In the additive manufacturing process, a number of internal vulnerabilities would manifest between successive printing layers that impact the internal microstructure and mechanics of the final products. These flaws include balling, porosity, fractures, powder aggregation, and thermal stress. Consequently, post-processing procedures are usually required to enhance the mechanical qualities and surface quality of the components after they are created and allow them to be used for the intended purpose (Peng et al., 2021).

> To obtain the required final properties, surface finish, and dimensional precision, additively manufactured metallic structures must undergo post-processing. It includes a variety of methods and procedures that are used following the first printing step. Numerous post-processing technologies are available to help conquer such issues, including thermal post-processing, Laser Shock Peening (LSP), Laser Polishing (LP), Conventional Machining Process (CMP), Milling, Rolling, Chemical Machining and Abrasive Machining, and Heating Processes (HP).
>
> (Mahmood et al., 2022; Airao et al., 2023)

This insight will facilitate the establishment of a research foundation and future projects related to this subject. This scientometric analysis is therefore necessary for the post-processing of structures made through additive manufacturing. Artificial intelligence (AI)-based scientometric analysis involves the analysis of scientific publications to identify and evaluate

 DOI: 10.1201/9781003506492-8

trends in a specific research area. In the case of post-processing of additively manufactured, a scientometric analysis would involve examining the relevant literature to gain insight into the research landscape, emerging trends, and major contributions in the field (Khang et al., 2023).

This chapter is organised as follows: A summary of the post-processing topic is given in Section 8.1. A comprehensive review of the pertinent research is given in Section 8.2. The material and study design are then covered in Section 8.3. The results of the systematic review, data analysis, and discussions regarding study trends are covered in detail in Section 8.4. Lastly, an overview of the systematic review and recommendations for additional research are provided in Section 8.5.

8.2 Associated Work

8.2.1 AI-Assisted Scientometric Analysis

AI systems can help academics with scientometric analysis, which is the measurement and study of scientific literature (Garfield 1979). In the long run, AI's ability to automatically analyse massive bibliometric data sets can aid in the speeding up of systematic reviews (Wang et al. 2022). However, if investigators and AI do not work harmoniously, the results may lack subject matter, precision, variation, and sufficient analysis (Correia and Lindley, 2022).

In 1969, Nalimov and Mulchenko invented the term "scientometrics" to describe "a measurement of science" (Zheng et al., 2021). Processing large volumes of bibliographies is challenging, especially when done manually; as a consequence, key links between diverse publications are periodically lost. As a result, this systematic approach becomes a crucial instrument for scholars. Given the volume of literature gathered, scientometric evaluation is a suitable initial step in locating and verifying the most pertinent data, so the systematic review offers an in-depth understanding of the results. To achieve this, a variety of techniques were used, such as authorship analysis, keyword co-occurrence, literature coupling analysis, and countries/regions activities (Castillo et al., 2022). The research clusters that have been identified offer valuable insights into the organisation of the knowledge realm related to additive manufacturing (AM) post-processing. However, scientometric analysis is limited in its capacity to present all research requests and obstacles. A systematic review was conducted to improve the scientometric evaluation and the research subject classification scheme that resulted from a consensus-based examination of the study's findings. Using a scientometric analytic approach, the current study examined the results of earlier research on post-processing in additive manufacturing.

8.2.2 Post-Processing of Additively Manufactured Metallic Structures

The AM technique involves the layering of material to create sophisticated 3D components from a specified model. AM is appealing for precise design and cost-effective quantities in industries, such as automotive, medicinal, and aerospace (Airao et al., 2023). It has been noticed that AM parts are not always acceptable for direct use in an application because they require additional processing to remove flaws, such as surface roughness, residual stresses, wear, porosity, and micro-hardness. Low surface qualities are linked to low mechanical properties, limiting component life. As a result, the AM metal components are post-processed to improve their mechanical characteristics and microstructure features. Both conventional (turning, milling, and drilling) and unconventional (laser machining, abrasive flow machining, and electrochemical polishing) post-processing techniques are used to reduce the flaws. The standard methods are more common than the non-conventional ones for post-processing because they restore surface quality, tolerances, and geometric precision (Airao et al., 2023).

8.3 Resources and Procedures

The resources accessed for data analysis and the research methodology for the study are elaborated in this section.

8.3.1 Extraction of Raw Data

Data collection from accessible sources was critical to the success of this scientometric analysis. This study's data gathering followed an approach for bibliometric evaluations. The authors applied two methods to narrow down the research on the discovered problem: (1) they investigated publications from 2013 to 2023 and carefully modified the literature through examining keywords and abstracts, and (2) they restricted themselves to peer-reviewed journal papers, which are reviewed for strict evaluations to remove errors and inaccuracies. Using the database was a crucial decision for the literature review. The authors opted to focus on Scopus instead of other resources as research repositories and academic databases offered extensive coverage of the topic in journal publications. The keywords selected were (post AND processing AND in AND additive AND manufacturing) in accordance with the research goal. The entire body of research on AM post-processing was located by searching the Scopus database for particular terms used in a publication's title, abstract, or keywords. To ascertain the current development of additively manufactured structure post-processing, the search window of 2013–2023 was chosen. The quantity of papers published in peer-reviewed English journals was limited through the use of a screening process.

8.3.2 Research Methods

A scientometric analysis uses visualisation-mapping tools and a research database to look for trends and findings in the research. Scientometrics is used in a variety of areas to evaluate the outcomes of publications, authors, journals, and research institutions. By using scientific citations, it is defined as offering a more thorough comprehension of research and a deeper understanding of scientific citations. In order to obtain a thorough grasp of the development of this field of study, the current endeavour aims to perform a scientometric evaluation and analysis of articles pertaining to the post-processing of additively generated structures from 2013 to 2023. The study examined the current state of research on post-processing in AM by combining systematic review and scientometric analysis into a mixed review methodology. This study's contribution is that it applies thorough research with a thorough quantitative evaluation and analysis using scientific network mapping methods, building on earlier review studies on the subject. On October 30, 2023, the documents were eliminated using an advance search query of:

> TITLE-ABS-KEY (*post* AND *processing* AND *in* AND *additive* AND *manufacturing*) AND (LIMIT-TO (DOCTYPE , *"ar"*)) AND (LIMIT-TO (LANGUAGE , *"English"*)) AND (LIMIT-TO (EXACTKEYWORD , *"Additive Manufacturing"*) OR LIMIT-TO (EXACTKEYWORD , *"Post-processing"*) OR LIMIT-TO (EXACTKEYWORD , *"Additive Manufacturing Process"*) OR LIMIT-TO (EXACTKEYWORD , *"Post-processing Techniques"*) OR LIMIT-TO (EXACTKEYWORD , *"Postprocessing Methods"*) OR LIMIT-TO (EXACTKEYWORD , *"Additive Manufacturing (AM)"*)) AND (LIMIT-TO (PUBYEAR , *2023*) OR LIMIT-TO (PUBYEAR , *2022*) OR LIMIT-TO (PUBYEAR , *2021*) OR LIMIT-TO (PUBYEAR , *2020*) OR LIMIT-TO (PUBYEAR , *2019*) OR LIMIT-TO (PUBYEAR , *2018*) OR LIMIT-TO (PUBYEAR , *2017*) OR LIMIT-TO (PUBYEAR , *2016*) OR LIMIT-TO (PUBYEAR , *2015*) OR LIMIT-TO (PUBYEAR , *2014*) OR LIMIT-TO (PUBYEAR , *2013*)).

8.3.3 Article Selection

A significant aspect of the meta-science evaluation carried out in this literature review is the method utilised to filter papers. Finding research themes, trends, and post-processing enhancements is one of the main goals of this analysis. As shown in Figure 8.1, the Scopus database was used for this review in order to accomplish this goal. Scopus was accessible through Punjab

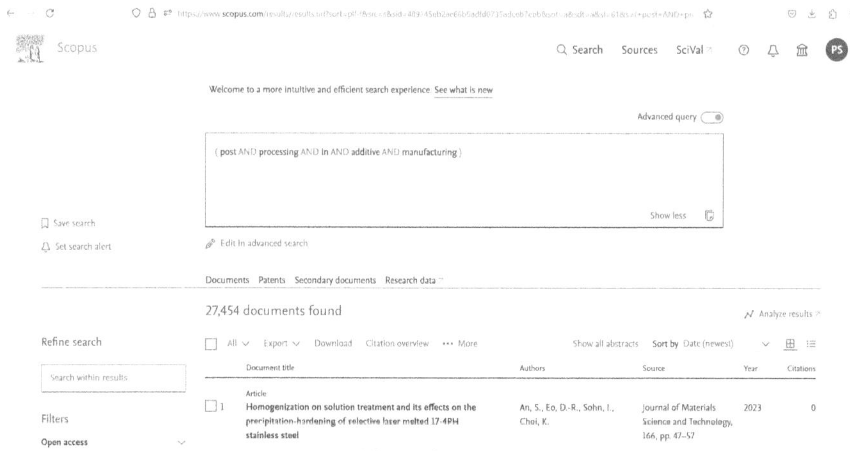

FIGURE 8.1
Keyword "post-processing in additive manufacturing" for meta-analysis. (*Source*: Scopus database.)

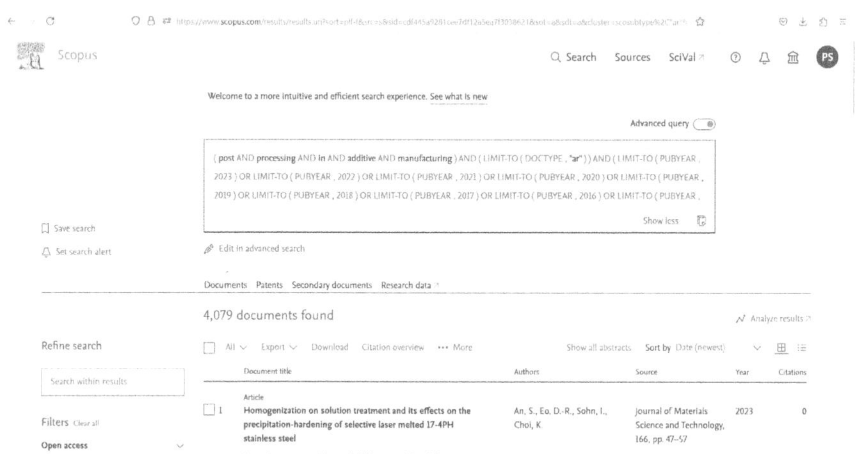

FIGURE 8.2
Presentation of data upon application of advance search.

University. Following precise modifications and exclusions to guarantee that the data used ought to reflect specialised on post-processing in AM, a total of 4,079 publications were taken into account in the meta-analysis. English descriptors were retrieved from the Scopus database (Figure 8.2).

8.3.4 The VOS Viewer

The VOSviewer, an open-source application, was used in this study to model and visualise networks (Amaechi et al., 2022).

> VOSviewer is an application for creating and visualizing bibliometric networks. These networks, which may include journals, researchers, or individual articles, can be formed via citation, bibliographic coupling, co-citation, or co-authorship links. VOSviewer also has text mining capabilities that allow you to create and visualize co-occurrence networks for significant terms retrieved from scientific literature. VOSviewer 1.6.19 appeared on January 23, 2023. This update improves VOSviewer's ability for creating maps using OpenAlex data.
>
> (Centre for Science and Technology Studies, Leiden University, The Netherlands, 2023)

8.3.5 R Programming

> R is a statistical computing and graphics language and framework. The S language and environment, created by John Chambers and associates at Bell Laboratories (formerly AT&T, now Lucent Technologies), served as the foundation for this GNU project. R can be thought of as a different method to carry out S. Despite significant variations, a large portion of the code created for S remains functional in R. A wide range of statistical (linear and nonlinear simulation, classical statistical tests, time-series evaluation, classification, and segmentation) and graphical tools are available in R, which is incredibly extensible. R is an open-source alternative to S, which is a well-liked computer language for statistical methods research.
>
> (Wei & Jiang, 2023)

8.4 Results and Analysis

8.4.1 Publication Background

The impact of research and a summary of the years of publication are included in the initial set of component systematic review results. Table 8.1 and Figure 8.3 indicate that the data was collected on 14 July 2023 from the Scopus database. When the most recent articles were included, the results showed a moderate shift in the publishing results for the years 2014 within 2022. The output shows a surge from 2014 in 2015. The availability of reliable additive manufacturing post-processing technology could be the cause of the recent surge in research. After that, it increased to 856 in 2021 and reached its peak of publications in 2022 (1,064 documents).

8.4.2 Origin of Literature

The primary focus of this study is the literary canon. However, other academic databases were looked through in order to verify the information from

TABLE 8.1

Publications vs the Years of Publication from 2013 to 2023

Year	No of Publications
2023	635
2022	1,064
2021	856
2020	567
2019	385
2018	274
2017	142
2016	82
2015	39
2014	13
2013	22

the Scopus database. Despite taking into account other important research databases like PubMed, Science Direct, DOAJ, Web of Science, and Google Scholar, the decision was made to use Scopus. Additional insights regarding the topic of post-processing of additively manufactured materials could be obtained through a thorough, methodical, and scientometric analysis of scholarly journal articles. Highly important specialised journals were also included (see Table 8.2), including *Additive Manufacturing*, which has an impact factor of 11.63 and a citescore of 17 (Figure 8.4).

The data presented in Table 8.4 and Figure 8.5 from Biblishiny indicates that the following journals published articles: *Additive Manufacturing* (371 articles), *Materials* (270 articles), *Materials Science and Engineering A* (217 articles), *International Journal of Advanced Manufacturing Technology* (209 articles), *Materials and Design* (126 articles), *Metals* (109 articles), *Rapid Prototyping Journal* (96 articles), and *International Journal of Fatigue* (92 articles). The *Journal of Materials Engineering and Performance* published 87 papers, while the *Journal of Manufacturing Processes* published 81 articles.

8.4.3 Publication Subjects

In this section the literature search using the published subjects displayed in Figure 8.6 is the main focus. These are the articles on the post-processing of additively created structures categorised according to their subjects. In the 2023 data, the largest percentage was found in engineering-related disciplines, accounting for 32.9%; *Materials Sciences* followed closely at 32.8%; together, these two disciplines made up more than 50% of the quadrat on publication themes. *Physics* and *Computer Science* came in second and third, respectively, followed by *Mathematics, Chemical Engineering, Chemistry, Business Management,* and least energy at 1.6%. Others, which included smaller

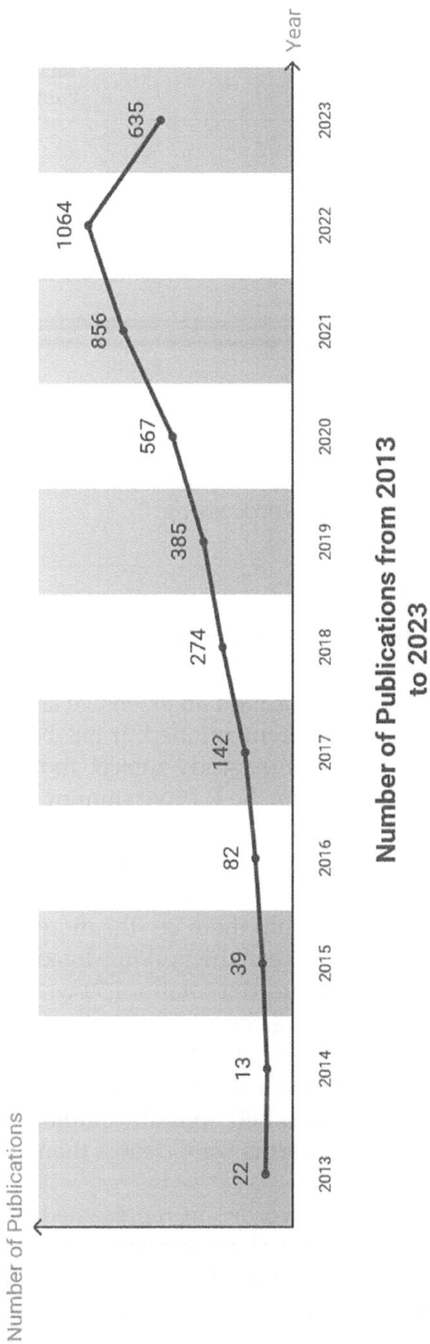

FIGURE 8.3
Graphical representation of year-wise growth pattern of publications from 2013 to 2023.

TABLE 8.2

Publications in the Top Journals for Post-Processing in Additive Manufacturing

Name of Journal	Total Publications	Impact Factor
Additive Manufacturing	371	11.632
Materials	270	3.748
Materials Science and Engineering: A	217	6.4
International Journal of Advanced Manufacturing Technology	209	3.4
Materials and Design	126	8.4
Metals	109	2.9
Rapid Prototyping Journal	96	4.043
International Journal of Fatigue	92	5.4
Journal of Materials Engineering and Performance	87	2.036
Journal of Manufacturing Processes	81	5.684

subgroups, came in at 3.8%, suggesting that there were other developing fields that performed on AM post-processing.

8.4.4 Publication Keywords

The study took place with a keyword search and sorting of data from various sources in the Scopus database. This is an example of keyword importance, and it is the first tool for sifting through an extensive amount of data on the research topic. In light of the aforementioned, it has been determined that keywords indicate both the literature study subject and how these ideas are related. To better visualise the data, the R Programming package Biblioshiny was used to display the connection between terms from diverse sources. The graph is simple to interpret: the gradients of the keywords are represented by the colours and sizes, and the distance between these keywords indicates how strong or close of a relationship there is—the more times a term occurs together, the stronger the link between them. While looking for co-occurrence of a keyword the first filter produced a total of 250 words that satisfied the criterion, with the aim of reducing the word count to facilitate easier visualisation in Biblioshiny. After examining the results, it emerged that, while the terms had identical significance, their writing styles distinguished them. The keywords were examined initially employing a word cloud, which demonstrated that some terms were denser than others, as shown in Figure 8.7.

The most densely packed keywords are represented by larger font sizes and distinct font colours. A word cloud generator is used to visualise the keywords in order, showing the highest to the lowest as boldest to least bold. The word cloud depicts that the most frequent author keywords include 3D printers, additives, selective laser melting, post-processing, and so on in Figure 8.7.

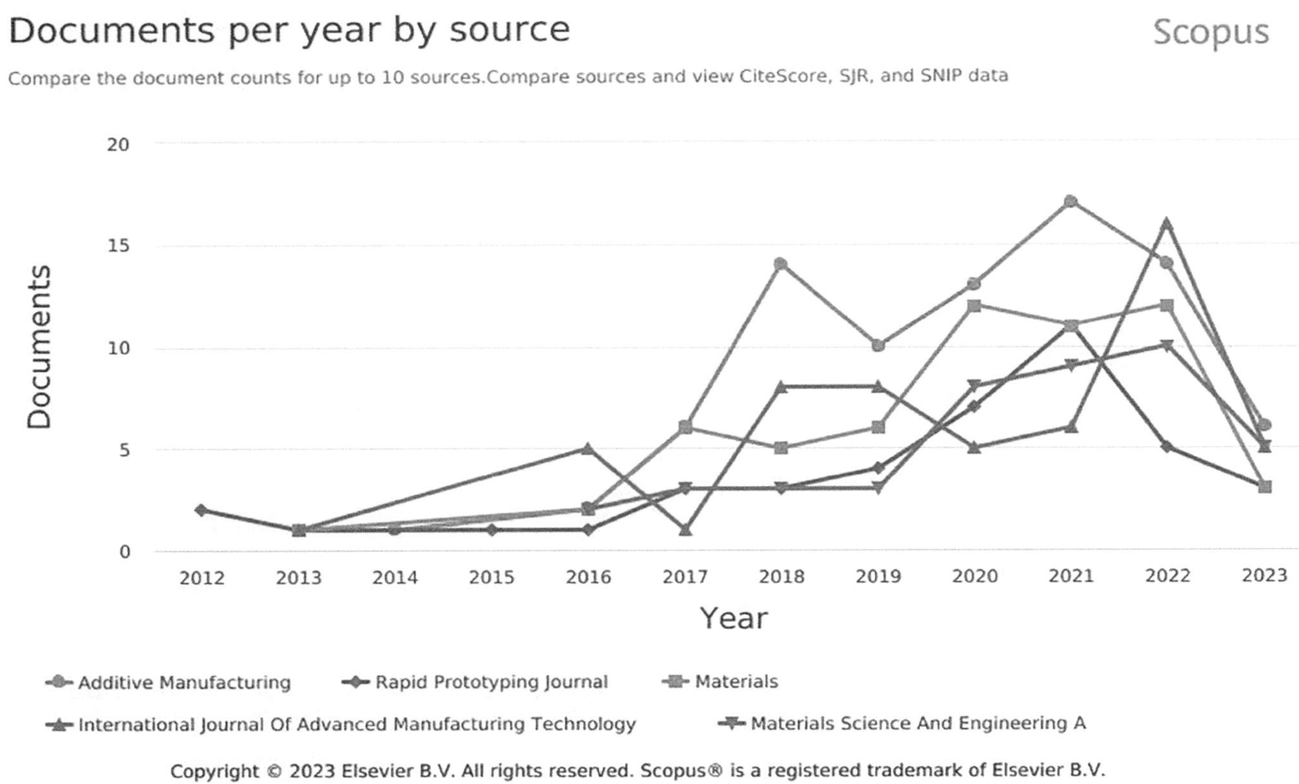

FIGURE 8.4
Graphical representation of various sources of literature and its year-wise growth pattern.

Publication Totals for Additive Manufacturing

Publication Venue	Total Publications
Additive Manufacturing	371
Materials	270
Materials Science and Engineering: A	217
International Journal of Advanced Manufacturing Technology	209
Materials and Design	126
Metals	109
Rapid Prototyping Journal	96
International Journal of Fatigue	92
Journal of Materials Engineering and Performance	87
Journal of Manufacturing Processes	81

FIGURE 8.5
Publications in the top journals for post-processing in additive manufacturing.

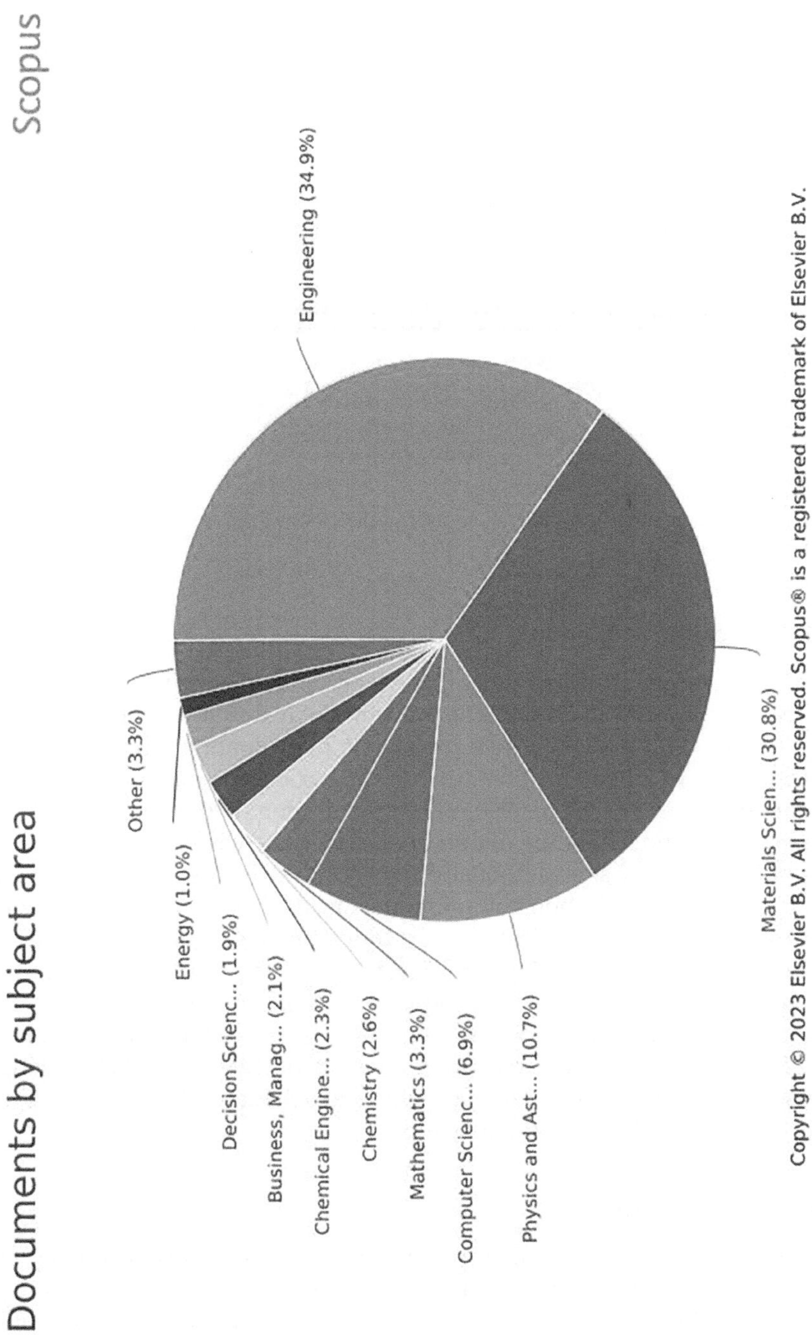

FIGURE 8.6
Distribution of literary text searches on subject-based publication classification.

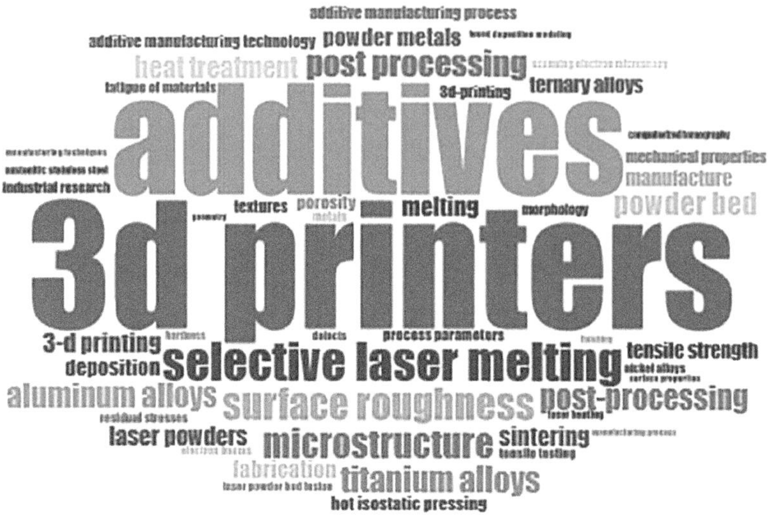

FIGURE 8.7
Word cloud based on published works about the research of "post-processing in additive manufacturing".

8.4.5 Publication Affiliation

This section presents the findings from the bibliometric analysis of the subject area's publication affiliation. The outcomes of research output connected to publications in this field play a major role in evaluating research trends and the impact of affiliations (institutions and organisations) on research. To calculate the impact of the institution or organisation on the investigation, a more thorough understanding of the support given by various affiliations for the post-processing assessment of additively built structures, which is given as an overview of publication volume from various departments, is required.

The most productive institutions are found in the United States, Asia, and Europe. Table 8.3 and Figure 8.8 summarise the top ten institutions. According to statistical research, these institutions published 30% of all documents. The Northwestern Polytechnical University takes first position with 186 publications, followed by the University of Waterloo and Huazhong University of Science and Technology, which take the second and third place, respectively, with 147 and 140 publications.

8.4.6 Publication Authors

A further part of the research trends analysis on "post-processing of additively produced metallic structures" will be carried out based on publication authorship. The impact of authorship on the study and the dissemination of publication volume is understood in the first section of the element

TABLE 8.3

Most Relevant Publication Affiliation for "Post-Processing in Additive Manufacturing" Research

Affiliating Institutions	Publications
Northwestern Polytechnical University	186
University of Waterloo	147
Huazhong University of Science and Technology	140
Politecnico Di Milano	124
Auburn University	120
Dalian University of Technology	113
Purdue University	99
University of New Brunswick	99
Shanghai Jiao Tong University	94

meta-analysis. Many researchers have contributed to the scientific literature on post-processing in AM. To illustrate the graph, the author names were then filtered so that they appeared only in articles with authors per publication. Figure 8.9 displays the authoring contributions for this topic. The collection of documents has over 2,000 authors, some of whom have only published one book. Wang Y has the greatest h-index, according to Table 8.4 from this study. In decreasing order of h-index, Zhang Y, Zhang Z, Li H, Wang H, Wang Z, Li J, Zhang X, Chen Y, and Liu Y are in order of h-index. It needs to be taken into account that the majority of recent works by authors with the greatest h-index are the ones that are most frequently mentioned in related works. Nonetheless, authors with the highest h-index tend to be those who have studied post-processing in enormous detail over an extended length of time.

8.4.7 Bradford Distributions of Articles

Bradford's law is a broad rule that controls how publications are distributed among journals in a specific scientific field. Samuel Bradford determined that a small number of core journals create around one-third of all articles in each area. The second zone of journals will also include around one-third of all publications, but N times as many journals. The third zone of journals will also include one-third of the articles, but in N journals. The number of journals in the core, second, and third zones has a mathematical connection of 1:N:N2 (Thompson & Walker, 2015).

Bradford law states that three zones can be identified, each of which generates one-third of all relevant papers, if numerous articles are arranged in reducing order of reducing effectiveness of journals (sources) pertinent to a given topic (Singh et al., 2007). According to Bradford law, Table 8.5 illustrates the ranking and zone distribution of articles. According to the results, *Additive Manufacturing* is ranked first out of ten journals, and seven journals fall into

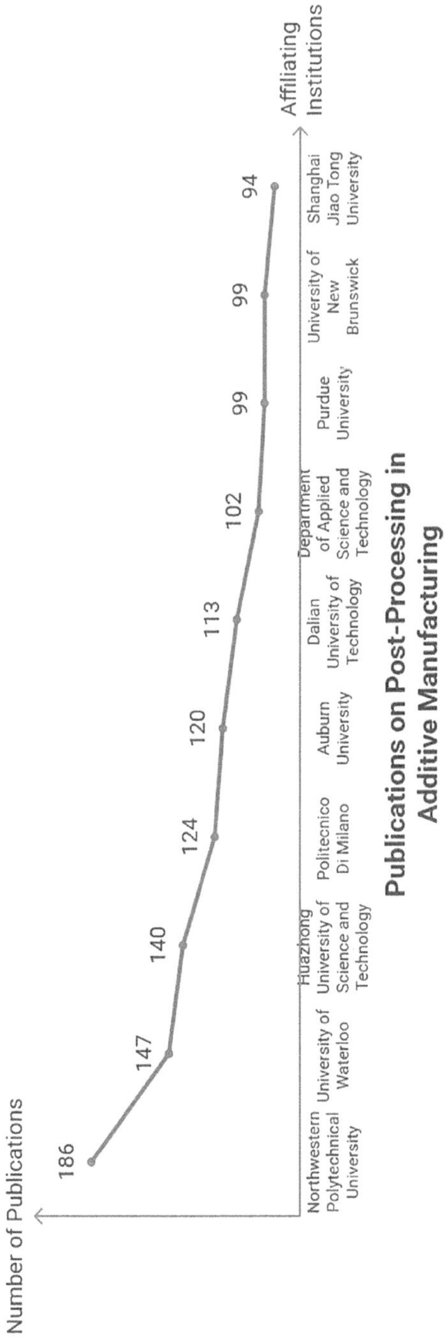

FIGURE 8.8
Most relevant publication affiliation.

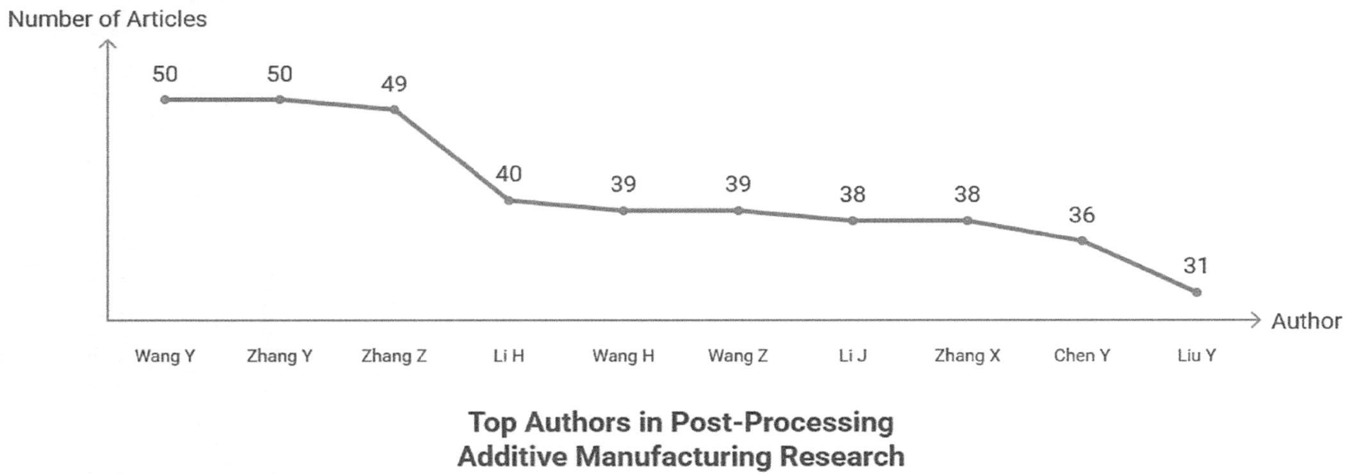

FIGURE 8.9
Most relevant publication authors.

TABLE 8.4

Most Relevant Top Ten Authors for "Post-Processing in Additive Manufacturing" Research

Authors	Articles	Articles Fractionalised
Wang Y.	50	9.68
Zhang Y.	50	10.27
Zhang Z.	49	8.97
Li H.	40	6.69
Wang H.	39	6.59
Wang Z.	39	8.17
Li J.	38	6.31
Zhang X.	38	6.41
Chen Y.	36	6.41
Liu Y.	31	5.19

TABLE 8.5

Ranking and Zone Distribution of Articles According to Bradford

Journals	Rank	Freq	Cumulative Freq	Zone
Additive Manufacturing	1	371	371	Zone 1
Materials	2	270	641	Zone 1
Materials Science and Engineering: A	3	217	858	Zone 1
International Journal of Advanced Manufacturing Technology	4	209	1067	Zone 1
Materials and Design	5	126	1193	Zone 1
Metals	6	109	1302	Zone 1
Rapid Prototyping Journal	7	96	1398	Zone 1
International Journal of Fatigue	8	92	1490	Zone 2
Journal of Materials Engineering and Performance	9	87	1577	Zone 2
Journal of Manufacturing Processes	10	81	1658	Zone 2

zone 1, indicating that these are the most productive journals, as shown in Figure 8.10.

8.4.8 Publication Countries/Regions

Figure 8.11 lists the top ten publishing countries, which include USA, China, Germany, Italy, UK, India, Australia, Canada, South Korea, and France. The USA ranks as the top country in terms of the total number of publications that are not the result of international collaboration. Scopus databases were found to represent near findings for each nation from 2013 to 2023, as shown in Table 8.6. A total of 4,286 publications were created in the USA according to the Scopus database. Germany, China, and the UK are also among the top four nations for scientific output. This makes it clear that research on the

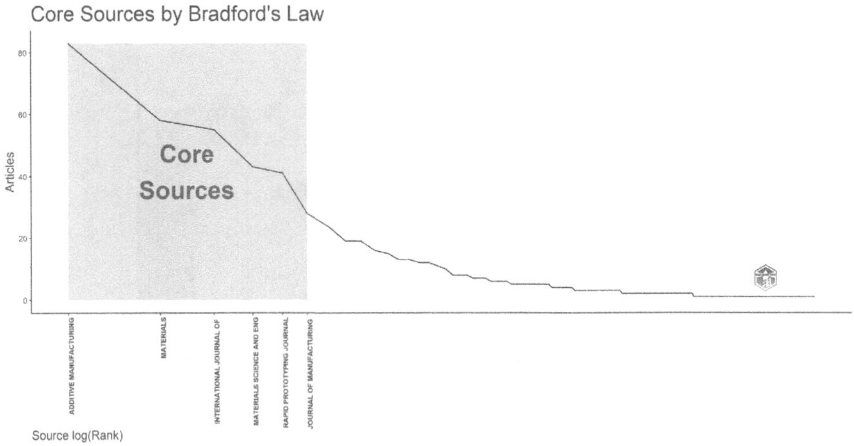

FIGURE 8.10
Bradford distributions of articles.

post-processing of materials used in additively created metallic structures is more popular in Europe and Asia, where there are more than 60 publications. As shown in Figure 8.11, the countries with the largest global network of collaboration, the USA, Germany, China, and the UK are at the leading edge of academic engagement.

Figure 8.12 depicts the evolution of country productivity over time, confirming the importance of the top three countries, the USA, China, and Germany, which have maintained a substantial rise since the beginning of the field of knowledge (2013). This examination of theme progression can be used as a technique to help academics comprehend the trend and knowledge gaps in the topic under consideration. Also, notice how the evolution of the major countries has accelerated rapidly at an exponential rate since 2018, compared to previous years. In this way, it is also found that research in post-processing AM is more popular in the European and Asian regions, where more than 50 articles have been published. As illustrated in Figure 8.13, academic engagement is led by the USA, China, and the UK, which have the world's largest collaboration network.

8.4.9 Future Research Trends in Post-Processing

The efficacy of post-processing treatments of an AM build cycle in final product requirements determines the absolute need of metal AM products for large targeted applications. On the other hand, the resilience of the constructed framework and surface integrity significantly impedes its widespread industrialisation. As a result, basic research that links additive and subtractive treatments leads to improvements in the performance and functioning of manufactured items. As a contemporary trending research, an

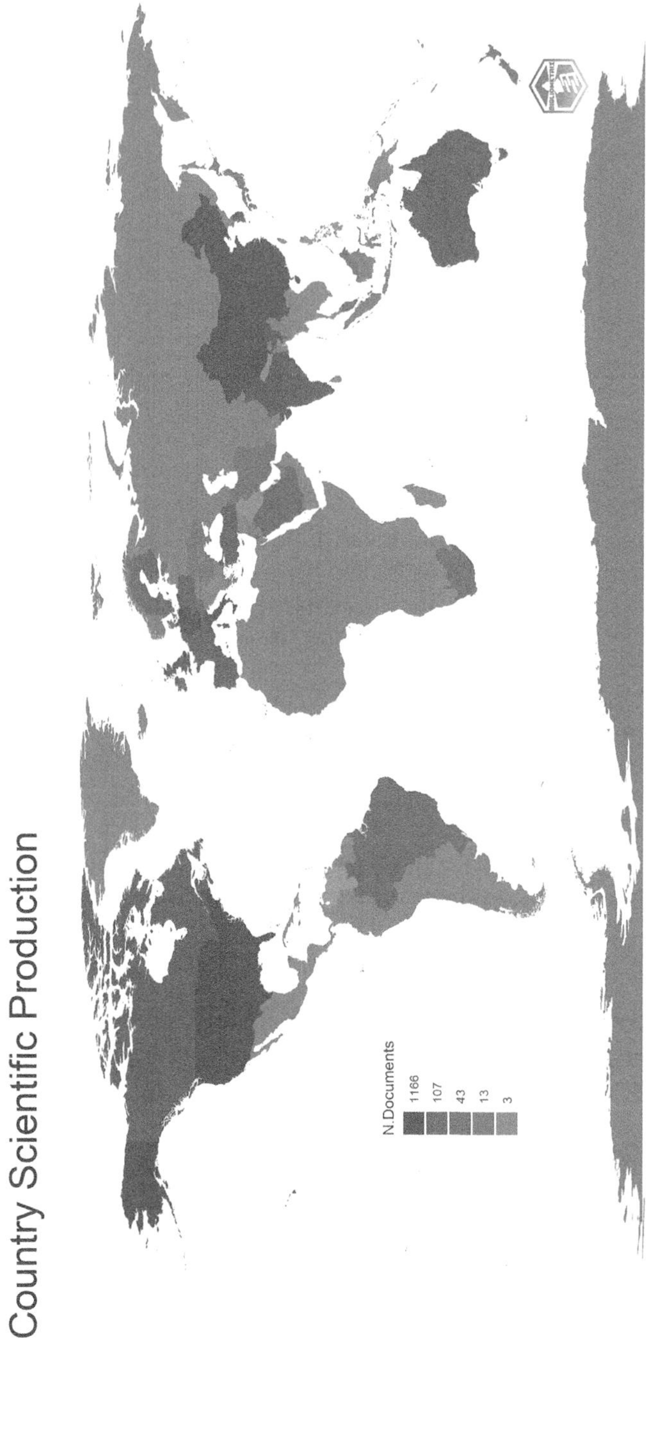

FIGURE 8.11
Collaboration network between the most relevant countries.

TABLE 8.6
Global Research Initiatives Highlight the Most Relevant Countries

Countries/Territories	Frequency of Documents
USA	4,286
China	2,939
Germany	1,916
Italy	1,230
Uk	1,158
India	877
Australia	757
Canada	701
South Korea	618
France	560

extensive area for investigation in the prediction of microstructural changes and grain growth prediction via numerical simulation and modelling is provided (S. et al., 2020).

Axinte and Billingha's (2019) physics-based research on adjusting process parameters causes microstructure changes influence on allowed stress attributes could be generalised to machinability assessment of additively made materials. The optimisation issue, of course, serves a significant role in decreasing time and effort by minimising material losses and selecting appropriate parametric configurations for layering processes in AM (Fayed et al., 2021). AM technology optimisation is often time expensive, particularly for systems that require extensive modelling. To address this, an inverse method of problem-solving based on managing parametric impacts on the build cycle could characterise microstructural alterations for any material and AM setup (Rubenchik et al., 2018). The major concern associated with fabricated AM products is that the surface roughness does not meet the precision requirements, and thereby it is unavoidable (Lee et al., 2022).

Surface finishing studies are scarce due to inadequate information and inappropriate experimental design (Melia et al., 2020). As a result, post-finishing procedures must be carried out with caution by designing appropriate surface roughness values. Conventional clenching, on the other hand, is a significant exception due to its ability to complete any freeform surface. However, this method has not acquired widespread acceptance for industrial applications (Yang et al., 2020). The challenging and tedious experimental setup design serves as a foundation for future research directions finding the Preston's law coefficient for a particular work piece material–tool pair. The research on the impact of post-processing on the mechanical, fatigue, and corrosion properties of AM-fabricated structures has only looked at just a handful of metals and alloys thus far. The microstructural characteristics, melt pool geometries, material inclusions, accumulated produced oxides, flaws, residual stresses, etc., are major determinants of these features (Levkulich

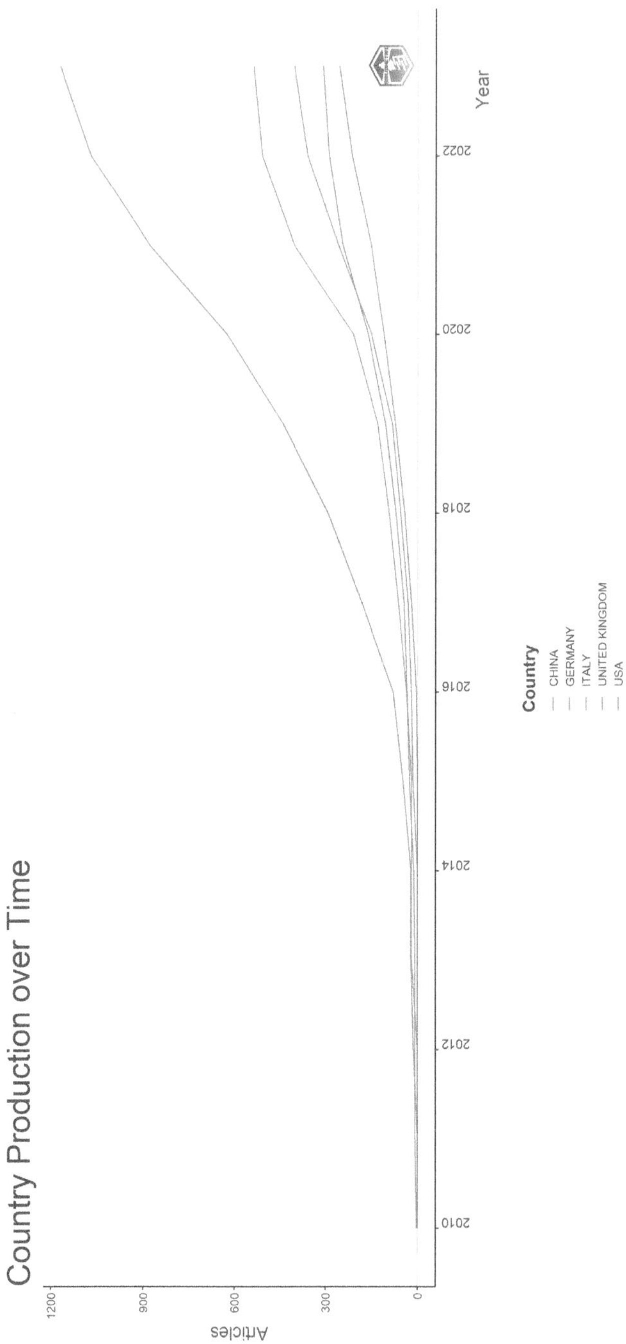

FIGURE 8.12
Evolution of country productivity over time.

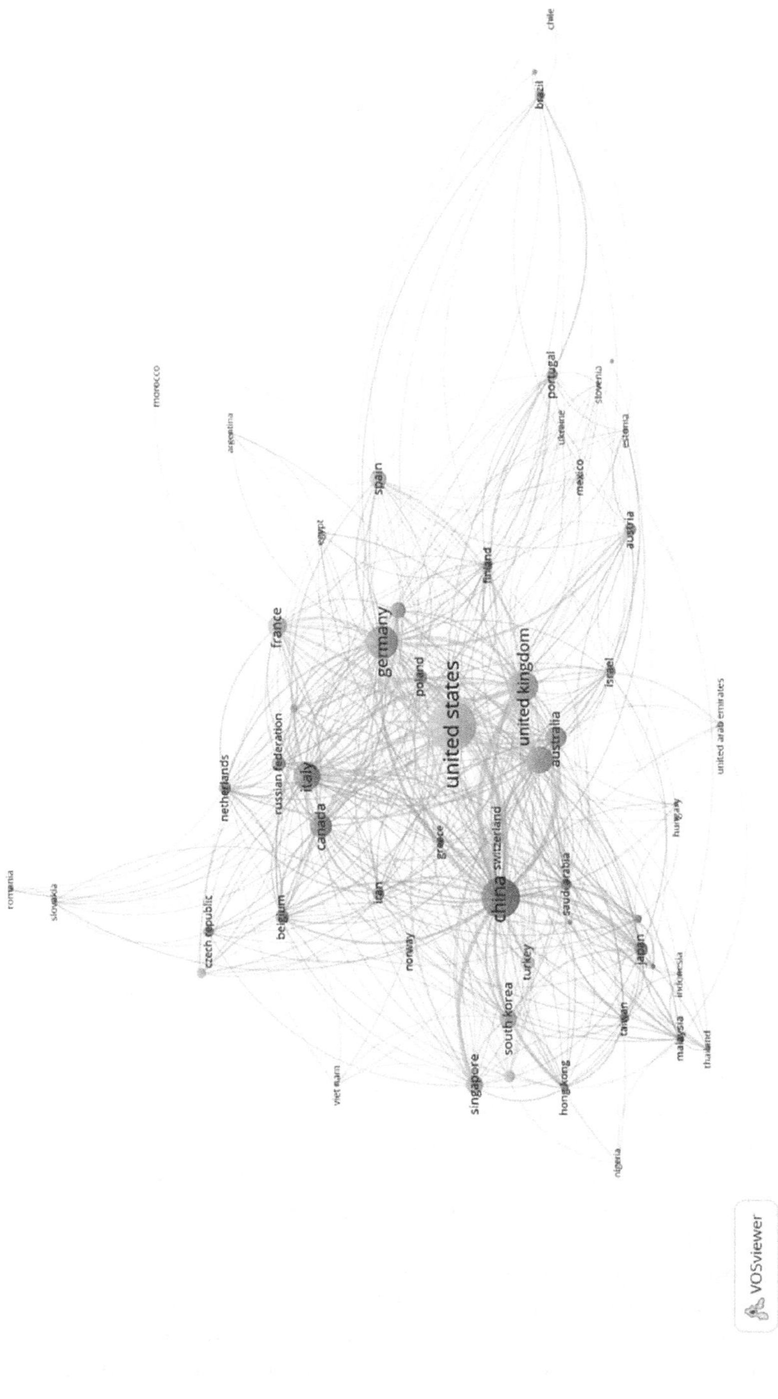

FIGURE 8.13
Collaboration network between the most relevant countries.

et al., 2019; Hae Young Lee et al., 2019). This makes it challenging to convey an overall evaluation of how post-processing affects the final product's usefulness. This indicates that the optimal design guideline for the AM process itself can be established, followed by post-processing and subtractive finishing treatments, to produce the improved functional qualities. The tribological and biological qualities are also supported by this argument, and it is believed that additional research will demonstrate the need for microstructural characteristics across a broad variety of material combinations, both before and after post-processing treatments. AM constructions need to have their build surface texture reduced via machining and/or finishing in order to fulfilling the current requirements for excellent formability, accuracy, and surface integrity. Furthermore, current research is focused on designing and developing equipment with the goal of producing same features and characteristics for identical material formulations.

8.5 Conclusions

This study uses AI to investigate the state of post-processing in AM at the moment. Comprehensive reviews of the literature in the field were carried out with the use of data from the WoS and Scopus databases. Subsequently, the primary research architectures, articles, and significant research themes were identified by evaluating publications, literature coupling, keyword co-occurrence, authorship, and countries/regions events. To achieve this goal, a comprehensive evaluation was conducted to identify best practices, and a scientific analysis was conducted to quantify the available data. Database such as Scopus, Vos-viewer, and R Programming (Biblioshyne) were also employed. This study provides a methodical evaluation of the most significant recent developments in AM post-processing, both in terms of methodology and applications, for works published from 2013 to 2023.

According to the scientometric analysis, there is a strong preference for investigating materials in this domain, with less emphasis on developing technology. There are various post-processing aspects that are pertinent to the subject area, based on the author keywords from this bibliometric review. According to this study, the USA was the most active region in the field of AM post-processing. This analysis also exposes writers' research preferences and affiliations. Expert focus groups can be included in other areas of scientific reviews, allowing papers to be classified into a variety of categories. The findings can be summarised as follows:

This research gives some important findings. First, the number of publications published increased considerably in 2015. In 2022, there is a significant increase in publication compared to previous years. The distribution of the selected articles by journal revealed that the majority of papers were published in the *Additive Manufacturing, Materials, Materials Science and Engineering A, International Journal of Advanced Manufacturing Technology, Materials and Design, Metals, Rapid Prototyping Journal,* and the *International Journal of Fatigue,* and in the USA, established and reputable researchers advance research in engineering-related disciplines. Similarly, Wang Y. is the author with the most publications, followed by Zhang Y. and Zhang Z. Furthermore, the most commonly used keywords by researchers were "3D printers", "additives", "selective laser melting", and "post-processing", and trends and research gaps in the study of post-processing are described.

References

Airao, J., Kishore, H., & Nirala, C. K. (2023). Comparative analysis of tool wear in micro-milling of wrought and selective laser melted Ti6Al4V. *Wear*, 523, 204788–204788. https://doi.org/10.1016/j.wear.2023.204788

Amaechi, C. V., Adefuye, E. F., Kgosiemang, I. M., Huang, B., & Amaechi, E. C. (2022). Scientometric review for research patterns on additive manufacturing of lattice structures. *Materials,* 15(15), 5323. https://doi.org/10.3390/ma15155323

Axinte, D., & Billingham, J. (2019). Time-dependent manufacturing processes lead to a new class of inverse problems. *Proceedings of the National Academy of Sciences of the United States of America,* 116(12), 5341–5343. https://doi.org/10.1073/pnas.1900420116

Castillo, M., Monroy, R., & Ahmad, R. (2022). Scientometric analysis and systematic review of smart manufacturing technologies applied to the 3D printing polymer material extrusion system. *Journal of Intelligent Manufacturing,* 35, 3–33. https://doi.org/10.1007/s10845-022-02049-1

Correia, A., & Lindley, S. (2022, December 1). Collaboration in relation to human-AI Systems: Status, trends, and impact. www.microsoft.com; www.microsoft.com/en-us/research/publication/collaboration-in-relation-to-human-ai-systems-status-trends-and-impact/

Fayed, E. M., Saadati, M., Shahriari, D., Brailovski, V., Jahazi, M., & Medraj, M. (2021). Optimization of the post-process heat treatment of Inconel 718 superalloy fabricated by laser powder bed fusion process. *Metals,* 11(1), 144. https://doi.org/10.3390/met11010144

Garfield, E. (1979). Scientometrics comes of age number46. https://garfield.library.upenn.edu/essays/v4p313y1979-80.pdf

Hae Young Lee, Han Jo Kim, Hong, H.-U., & Lee, B.-S. (2019). Influence of the focus offset on the defects, microstructure, and mechanical properties of an Inconel 718 superalloy fabricated by electron beam additive manufacturing. *Journal of Alloys and Compounds,* 781, 842–856. https://doi.org/10.1016/j.jallcom.2018.12.070

Khang, A., Vugar Abdullayev, Jadhav, B., Gupta, S., & Morris, G. (2023). AI-centric modeling and analytics. In *AI-Centric Modeling and Analytics: Concepts, Technologies, and Applications.* CRC Press eBooks. Informa. https://doi.org/10.1201/9781003400110

Lee, S.-J., Shao, S., Wells, D., Miroslav Zetek, Kepka, M., & Nima Shamsaei. (2022). Fatigue behavior and modeling of additively manufactured IN718: The effect of surface treatments and surface measurement techniques. *Journal of Materials Processing Technology*, 302(117475), 117475–117475. https://doi.org/10.1016/j.jmatprotec.2021.117475

Levkulich, N. C., Semiatin, S. L., Gockel, J. E., Middendorf, J. R., DeWald, A. T., & Klingbeil, N. W. (2019). The effect of process parameters on residual stress evolution and distortion in the laser powder bed fusion of Ti-6Al-4V. *Additive Manufacturing*, 28, 475–484. https://doi.org/10.1016/j.addma.2019.05.015

Mahmood, M. A., Chioibasu, D., Ur Rehman, A., Mihai, S., & Popescu, A. C. (2022). Post-processing techniques to enhance the quality of metallic parts produced by additive manufacturing. *Metals*, 12(1), 77. https://doi.org/10.3390/met12010077

Melia, M. T., Duran, J. G., Joshua Robert Koepke, Saiz, D. J., Bradley Howell Jared, & Schindelholz, E. J. (2020). How build angle and post-processing impact roughness and corrosion of additively manufactured 316L stainless steel. *Npj Materials Degradation*, 4(1). https://doi.org/10.1038/s41529-020-00126-5

Peng, X., Kong, L., Fuh, J. Y. H., & Wang, H. (2021). a review of post-processing technologies in additive manufacturing. *Journal of Manufacturing and Materials Processing*, 5(2), 38. https://doi.org/10.3390/jmmp5020038

Razanica, S., Malakizadi, A., Larsson, R., Cedergren, S., & Josefson, B. L. (2020). FE modeling and simulation of machining Alloy 718 based on ductile continuum damage. *International Journal of Mechanical Sciences*, 171, 105375. https://doi.org/10.1016/j.ijmecsci.2019.105375

Rubenchik, A. M., King, W. E., & Wu, S. S. (2018). Scaling laws for the additive manufacturing. *Journal of Materials Processing Technology*, 257, 234–243. https://doi.org/10.1016/j.jmatprotec.2018.02.034

Singh, G., Mittal, R., & Ahmad, M. (2007). A bibliometric study of literature on digital libraries. *The Electronic Library*, 25(3), 342–348. https://doi.org/10.1108/02640470710754841

Thompson, D. F., & Walker, C. K. (2015). A descriptive and historical review of bibliometrics with applications to medical sciences. *Pharmacotherapy: The Journal of Human Pharmacology and Drug Therapy*, 35(6), 551–559. https://doi.org/10.1002/phar.1586

VOSviewer. (2023). VOSviewer – Visualizing scientific landscapes. VOSviewer. www.vosviewer.com/

Wang, W., Jiang, X., Tian, S., Liu, P., Dang, D., Su, Y., Lookman, T., & Xie, J. (2022). Automated pipeline for superalloy data by text mining. *Npj Computational Materials*, 8(1). https://doi.org/10.1038/s41524-021-00687-2

Wei, W., & Jiang, Z. (2023). A bibliometrix-based visualization analysis of international studies on conversations of people with aphasia: Present and prospects. *Heliyon*, 9(6), e16839–e16839. https://doi.org/10.1016/j.heliyon.2023.e16839

Yang, Y., Li, H., Liao, Z., Axinte, D., Zhu, W.-L., & Beaucamp, A. (2020). Controlling of compliant grinding for low-rigidity components. *International Journal of Machine Tools & Manufacture*, 152(103543), 103543–103543. https://doi.org/10.1016/j.ijmachtools.2020.103543

Zheng, Yufan, et al. (2021) Scientometric analysis and systematic review of multi-material additive manufacturing of polymers. *Polymers*, 13(12), 1957, https://doi.org/10.3390/polym13121957.

Index